Textile Manufacturing: Processes and Techniques

Textile Manufacturing: Processes and Techniques

Edited by
Eli Lewis

Larsen & Keller
www.larsen-keller.com

Textile Manufacturing: Processes and Techniques
Edited by Eli Lewis
ISBN: 978-1-63549-274-3 (Hardback)

© 2017 Larsen & Keller

☰ Larsen & Keller

Published by Larsen and Keller Education,
5 Penn Plaza,
19th Floor,
New York, NY 10001, USA

Cataloging-in-Publication Data

Textile manufacturing : processes and techniques / edited by Eli Lewis.
    p. cm.
Includes bibliographical references and index.
ISBN 978-1-63549-274-3
1. Textile fabrics. 2. Textile industry. 3. Textile industry--Technological innovations.
4. Spinning. 5. Knitting. I. Lewis, Eli.
TS1765 .T49 2017
677--dc23

The publisher's policy is to use permanent paper from mills that operate a sustainable forestry policy. Furthermore, the publisher ensures that the text paper and cover boards used have met acceptable environmental accreditation standards.

Printed and bound in the United States of America.

For more information regarding Larsen and Keller Education and its products, please visit the publisher's website www.larsen-keller.com

Table of Contents

Preface

Textile manufacturing is a vast industry which is concerned with turning fiber into yarn and then yarn into fabric. It encompasses the machinery and methodology related to textile manufacturing. The stages included in textile manufacturing include cultivating and harvesting, preparatory processes, spinning, weaving, finishing and lastly marketing. This book is a valuable compilation of topics, ranging from the basic to the most complex theories and principles in the field of textile manufacturing. The topics included in it are of utmost significance and bound to provide incredible insights to readers. The various sub-fields of textile engineering along with technological progress that have future implications are glanced at in this textbook. It is an essential guide for both academicians and those who wish to pursue this discipline further.

A foreword of all chapters of the book is provided below:

Chapter 1 - The textile industry is massive and is responsible for the conversion of yarn to clothing. It comprises of various interconnected processes that transform raw materials into the finished product. This chapter introduces the reader to textile manufacturing by exploring the various variable processes and pre-industrial methods involved to generate the finished product.; **Chapter 2** - This chapter focuses on the various processes employed on cotton fabric in its loom state to remove impurities and to improve its quality. The reader is introduced to processes like desizing, scouring, bleaching, mercerizing, singeing, raising, calendering, sanforizing, dyeing and textile printing. The chapter studies these multiple facets of textile processing in-depth and provides the reader with comprehensive information.; **Chapter 3** - This chapter arms the reader with information regarding the processes that convert yarn to cloth or fabric. Weaving, knitting and crocheting have been studied in detail in the chapter. Such methods have been practiced for thousands of years. The chapter also talks about the machinery involved in these processes.; **Chapter 4** - Spinning refers to the process of the conversion of cotton to yarn. This chapter gives an insight into both the olden and contemporary methods of spinning like mule spinning, ring spinning and break or open-end spinning. A section of this chapter is also dedicated to the array of machinery used in spinning like the spinning wheel, water frame, spinning mule, spinning jenny, throstle, ring frame and dref friction spinning. The major components of textile spinning are discussed in this chapter.; **Chapter 5** - This chapter deals with the process of knitting and provides the reader with a comprehensive guide to the topic of knitting while detailing the different types of knitting like warp knitting, cable knitting, lace knitting and plaited knitting. The various types of knitting produce fabric of varying texture, stretch and symmetry. Textile manufacturing is best understood in confluence with the major topics listed in the following chapter.; **Chapter 6** - The revolutionary field of e-textiles integrates technology into fabric by producing garments that have digital components embedded into them. This chapter introduces the reader to the topic of electronic textiles or smart textiles, its history and categories. The various applications of these textiles and its future have also been explored in this chapter.; **Chapter 7** - Textiles find use and application in various fields depending on their properties like finish, texture, abilities (fire retardancy, water- proofing, thermal insulation etc.) stretch and the like. This chapter aims at providing the reader a thorough understanding of the various types of specialized fabrics like Gore-Tex, geotextile, microfiber and spandex.; **Chapter 8** - This chapter details processes of lace making and embroidery. These require special processes and techniques for manufacture. The chapter also talks about different types of lace like chemical lace, bobbin lace and needle lace. The section on embroidery focuses on machine embroidery.

At the end, I would like to thank all the people associated with this book devoting their precious time and providing their valuable contributions to this book. I would also like to express my gratitude to my fellow colleagues who encouraged me throughout the process.

Editor

Introduction to Textile Manufacturing

The textile industry is massive and is responsible for the conversion of yarn to clothing. It comprises of various interconnected processes that transform raw materials into the finished product. This chapter introduces the reader to textile manufacturing by exploring the various variable processes and pre-industrial methods involved to generate the finished product.

Textile Manufacturing

Textile manufacturing is a major industry. It is based on the conversion of fiber into yarn, yarn into fabric. These are then dyed or printed, fabricated into clothes. Different types of fiber are used to produce yarn. Cotton remains the most important natural fibre, so is treated in depth. There are many variable processes available at the spinning and fabric-forming stages coupled with the complexities of the finishing and colouration processes to the production of a wide ranges of products. There remains a large industry that uses hand techniques to achieve the same results.

Processing of Cotton

Cotton Manufacturing Processes				
Bale Breaker				Blowing Room
Willowing				
Breaker Scutcher		Batting		
Finishing Scutcher		Lapping		
Carding				Carding Room
Sliver Lap				

Combing			
Drawing			
Slubbing			
Intermediate			
Roving	Fine Roving		
Mule Spinning	Ring Spinning	Spinning	
	Reeling		Doubling
Winding	Bundling		Bleaching
Weaving shed			Winding
Beaming			Cabling
Warping			Gassing
Sizing/Slashing/Dressing			Spooling
Weaving			
Cloth	Yarn (Cheese)- - Bundle		Sewing Thread

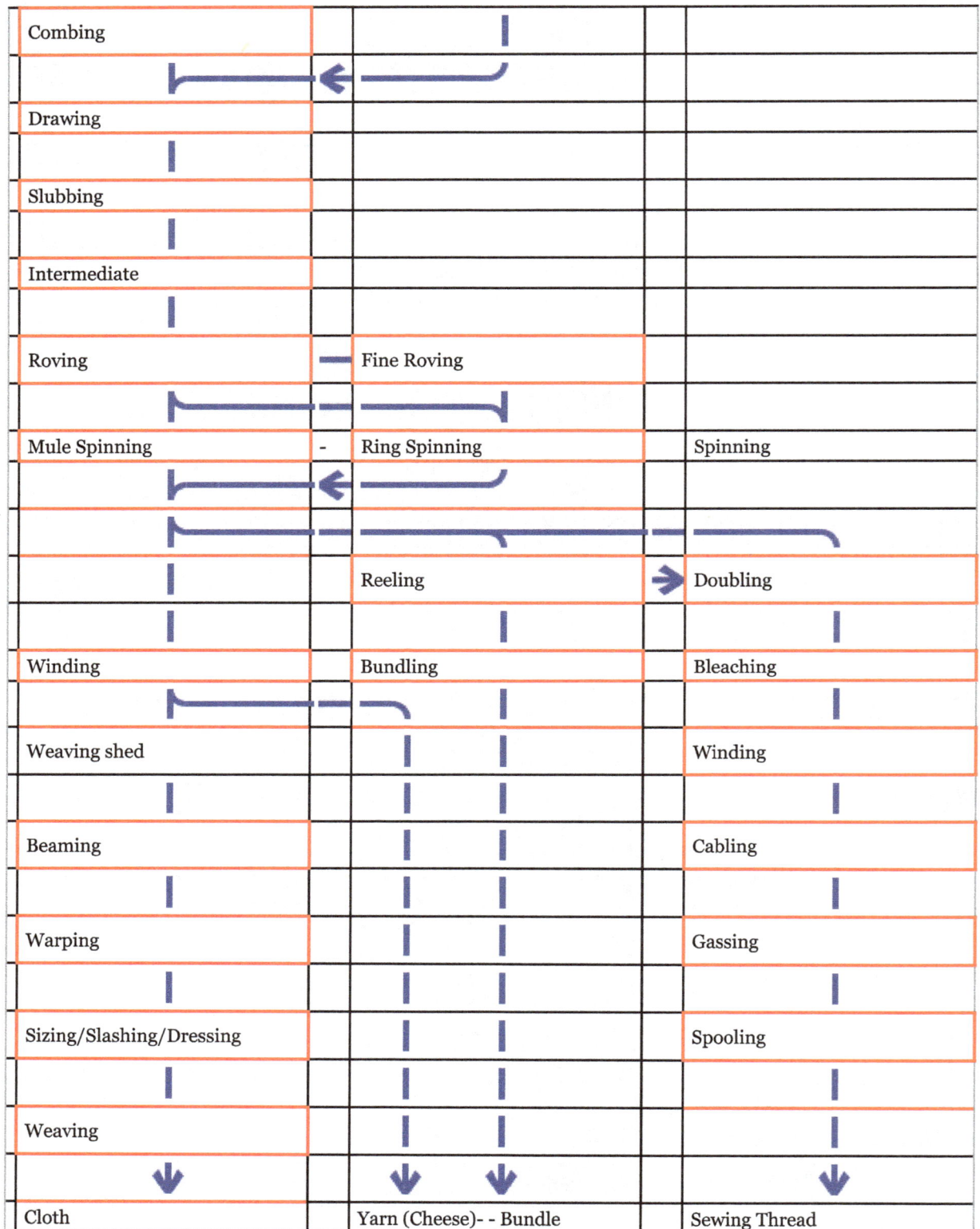

Cotton is the world's most important natural fibre. In the year 2007, the global yield was 25 million tons from 35 million hectares cultivated in more than 50 countries.

There are six stages

- Cultivating and Harvesting
- Preparatory Processes
- Spinning
- Weaving or Knitting
- Finishing
- Marketing

Cultivating and Harvesting

Cotton is grown anywhere with long, hot dry summers with plenty of sunshine and low humidity. Indian cotton, gossypium arboreum, is finer but the staple is only suitable for hand processing. American cotton, gossypium hirsutum, produces the longer staple needed for machine production. Planting is from September to mid November and the crop is harvested between March and June. The cotton bolls are harvested by stripper harvesters and spindle pickers, that remove the entire boll from the plant. The cotton boll is the seed pod of the cotton plant, attached to each of the thousands of seeds are fibres about 2.5 cm long.

Ginning

The seed cotton goes in to a Cotton gin. The cotton gin separates seeds and removes the "trash" (dirt, stems and leaves) from the fibre. In a saw gin, circular saws grab the fibre and pull it through a grating that is too narrow for the seeds to pass. A roller gin is used with longer staple cotton. Here a leather roller captures the cotton. A knife blade, set close to the roller, detaches the seeds by drawing them through teeth in circular saws and revolving brushes which clean them away.

The ginned cotton fibre, known as lint, is then compressed into bales which are about 1.5 m tall and weigh almost 220 kg. Only 33% of the crop is usable lint. Commercial cotton is priced by quality, and that broadly relates to the average length of the staple, and the variety of the plant. Longer staple cotton (2½ in to 1¼ in) is called Egyptian, medium staple (1¼ in to ¾ in) is called American upland and short staple (less than ¾ in) is called Indian.

The cotton seed is pressed into a cooking oil. The husks and meal are processed into animal feed, and the stems into paper.

Preparatory Processes - Preparation of Yarn

- Ginning, bale-making and transportation is done in the country of origin.
- Opening and cleaning

Cotton mills get the cotton shipped to them in large, 500 pound bales. When the cotton comes out of a bale, it is all packed together and still contains vegetable matter. The bale is broken open using a machine with large spikes. It is called an Opener. In order to fluff up the cotton and remove the vegetable matter, the cotton is sent through a picker, or similar machines. The cotton is fed into a machine known as a picker, and gets beaten with a beater bar in order to loosen it up. It is fed through various rollers, which serve to remove the vegetable matter. The cotton, aided by fans, then collects on a screen and gets fed through more rollers till it emerges as a continuous soft fleecy sheet, known as a lap.

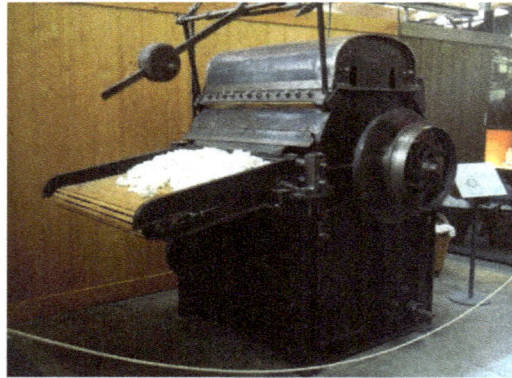

Platt Bros. Picker

Blending,

Mixing & Scutching

Scutching refers to the process of cleaning cotton of its seeds and other impurities. The first scutching machine was invented in 1797, but did not come into further mainstream use until after 1808 or 1809, when it was introduced and used in Manchester, England. By 1816, it had become generally adopted. The scutching machine worked by passing the cotton through a pair of rollers, and then striking it with iron or steel bars called beater bars or beaters. The beaters, which turn very quickly, strike the cotton hard and knock the seeds out. This process is done over a series of parallel bars so as to allow the seeds to fall through. At the same time, air is blown across the bars, which carries the cotton into a cotton chamber.

Carding

Carding: the fibres are separated and then assembled into a loose strand (sliver or tow) at the conclusion of this stage.

Carding machine

A Combing machine

The cotton comes off of the picking machine in laps, and is then taken to carding machines. The carders line up the fibres nicely to make them easier to spin. The carding machine consists mainly of one big roller with smaller ones surrounding it. All of the rollers are covered in small teeth, and as the cotton progresses further on the teeth get finer (i.e. closer together). The cotton leaves the carding machine in the form of a sliver; a large rope of fibres.

Note: In a wider sense Carding can refer to these four processes: Willowing- loosening the fibres; Lapping- removing the dust to create a flat sheet or lap of cotton; Carding- combing the tangled lap into a thick rope of 1/2 in in diameter, a sliver; and Drawing- where a drawing frame combines 4 slivers into one- repeated for increased quality.

- Combing is optional, but is used to remove the shorter fibres, creating a stronger yarn.

- Drawing the fibres are straightened

  Several slivers are combined. Each sliver will have thin and thick spots, and by combining several slivers together a more consistent size can be reached. Since combining several slivers produces a very thick rope of cotton fibres, directly after being combined the slivers are separated into rovings. These rovings (or slubbings) are then what are used in the spinning process.

  Generally speaking, for machine processing, a roving is about the width of a pencil.

  - Drawing frame: Draws the strand out

  - Slubbing Frame: adds twist, and winds onto bobbins

  - Intermediate Frames: are used to repeat the slubbing process to produce a finer yarn.

  - Roving frames: reduces to a finer thread, gives more twist, makes more regular and even in thickness, and winds onto a smaller tube.

Spinning - yarn Manufacture

- Spinning

Most spinning today is done using Break or Open-end spinning, this is a technique where the staples are blown by air into a rotating drum, where they attach themselves to the tail of formed yarn that is continually being drawn out of the chamber. Other methods of break spinning use needles and electrostatic forces. This method has replaced the older methods of ring and mule spinning. It also is easily adapted for artificial fibres.

The spinning machines takes the roving, thins it and twists it, creating yarn which it winds onto a bobbin.

In mule spinning the roving is pulled off a bobbin and fed through some rollers, which are feeding at several different speeds. This thins the roving at a consistent rate. If the roving was not a consistent size, then this step could cause a break in the yarn, or could jam the machine. The yarn is twisted through the spinning of the bobbin as the carriage moves out, and is rolled onto a cylinder called a spindle, which then produces a cone-shaped bundle of fibres known as a "cop", as the carriage returns. Mule spinning produces a finer thread than the less skilled ring spinning.

- The mule was an intermittent process, as the frame advanced and returned a distance of 5ft.It was the descendant of 1779 Crompton device. It produces a softer less twisted thread that was favoured for fines and for weft.

- The ring was a descendant of the Arkwright water Frame 1769. It was a continuous process, the yarn was coarser, had a greater twist and was stronger so was suited to be warp. Ring spinning is slow due to the distance the thread must pass around the ring, other methods have been introduced.

Sewing thread, was made of several threads twisted together, or doubled.

Checking.

This is the process where each of the bobbins is rewound to give a tighter bobbin.

Folding and Twisting

Plying is done by pulling yarn from two or more bobbins and twisting it together, in the opposite direction that in which it was spun. Depending on the weight desired, the cotton may or may not be plied, and the number of strands twisted together varies.

Gassing

Gassing is the process of passing yarn, as distinct from fabric very rapidly through a series of Bunsen gas flames in a gassing frame, in order to burn off the projecting fibres and make the thread round and smooth and also brighter. Only the better qualities of yarn are gassed, such as that used for voiles, poplins, venetians, gabardines, many Egyptian qualities, etc. There is a loss of weight in gassing, which varies' about 5 to 8 per cent., so that if a 2/60's yarn is required 2/56's would be used. The gassed yarn is darker in shade afterwards, but should not be scorched.

Mule spinning

Mule spinning

Ring spinning

Ring spinning

Measurements

- Cotton Counts: The number of pieces of thread, 840 yards long needed to make up 1 lb weight. 10 count cotton means that 10x840 yd weighs 1 lb. This is coarser than 40 count cotton where 40x840 yards are needed. In the United Kingdom, Counts to 40s are coarse

(Oldham Counts), 40 to 80s are medium counts and above 80 is a fine count. In the United States ones to 20s are coarse counts.

- Hank: A length of 7 leas or 840 yards

- Thread: A length of 54 in (the circumference of a warp beam)

- Bundle: Usually 10 lb

- Lea: A length of 80 threads or 120 yards

- Denier: this is an alternative method. It is defined as a number that is equivalent to the weight in grams of 9000m of a single yarn. 15 denier is finer than 30 denier.

- Tex: is the weight in grams of 1 km of yarn.

The worsted hank is only 560 yd

Weaving-fabric Manufacture

The weaving process uses a loom. The lengthway threads are known as the warp, and the cross way threads are known as the weft. The warp which must be strong needs to be presented to loom on a warp beam. The weft passes across the loom in a shuttle, that carries the yarn on a pirn. These pirns are automatically changed by the loom. Thus, the yarn needs to be wrapped onto a beam, and onto pirns before weaving can commence.

Winding

After being spun and plied, the cotton thread is taken to a warping room where the winding machine takes the required length of yarn and winds it onto warpers bobbins

Warping or Beaming

A Warper

Racks of bobbins are set up to hold the thread while it is rolled onto the warp bar of a loom. Because the thread is fine, often three of these would be combined to get the desired thread count.

Sizing

Slasher sizing machine needed for strengthening the warp by adding starch to reduce breakage of the yarns.

Drawing in, Looming

The process of drawing each end of the warp separately through the dents of the reed and the eyes of the healds, in the order indicated by the draft.

Pirning (Processing the Weft)

Pirn winding frame was used to transfer the weft from cheeses of yarn onto the pirns that would fit into the shuttle

Weaving

At this point, the thread is woven. Depending on the era, one person could manage anywhere from 3 to 100 machines. In the mid nineteenth century, four was the standard number. A skilled weaver in 1925 would run 6 Lancashire Looms. As time progressed new mechanisms were added that stopped the loom any time something went wrong. The mechanisms checked for such things as a broken warp thread, broken weft thread, the shuttle going straight across, and if the shuttle was empty. Forty of these Northrop Looms or automatic looms could be operated by one skilled worker.

A Draper loom in textile museum, Lowell, Massachusetts

The three primary movements of a loom are shedding, picking, and beating-up.

- *Shedding*: The operation of dividing the warp into two lines, so that the shuttle can pass between these lines. There are two general kinds of sheds-"open" and "closed." Open Shed-The warp threads are moved when the pattern requires it-from one line to the other. Closed Shed-The warp threads are all placed level in one line after each pick.

- *Picking:*The operation of projecting the shuttle from side to side of the loom through the division in the warp threads. This is done by the overpick or underpick motions. The overpick is suitable for quick-running looms, whereas the underpick is best for heavy or slow looms.

- *Beating-up*: The third primary movement of the loom when making cloth, and is the action of the reed as it drives each pick of weft to the fell of the cloth.

The Lancashire Loom was the first semi-automatic loom. Jacquard looms and Dobby looms are looms that have sophisticated methods of shedding. They may be separate looms, or mechanisms added to a plain loom. A Northrop Loom was fully automatic and was mass produced between 1909 and the mid-1960s. Modern looms run faster and do not use a shuttle: there are air jet looms, water jet looms and rapier looms.

Measurements

- Ends and Picks: Picks refer to the weft, ends refer to the warp. The coarseness of the cloth can be expressed as the number of picks and ends per quarter inch square, or per inch square. Ends is always written first. For example: *Heavy domestics are made from coarse yarns, such as 10's to 14's warp and weft, and about 48 ends and 52 picks.*

Associated Job Titles

- Piecer
- Scavenger
- Weaver
- Tackler
- Draw boy
- Pirner

Issues

When a hand loom was located in the home, children helped with the weaving process from an early age. Piecing needs dexterity, and a child can be as productive as an adult. When weaving moves from the home to the mill, children are often allowed to *help* their older sisters, and laws have to be made to prevent child labour becoming established.

Knitting- fabric Manufacture

Knitting by machine is done in two different ways; warp and weft. Weft knitting (as seen in the pictures) is similar in method to hand knitting with stitches all connected to each other horizontally. Various weft machines can be configured to produce textiles from a single spool of yarn or multiple spools depending on the size of the machine cylinder (where the needles are bedded). In a warp knit there are many pieces of yarn and there are vertical chains, zigzagged together by crossing the yarn.

A circular knitting machine.

Close-up on the needles.

Cotton

Warp knits do not stretch as much as a weft knit, and it is run-resistant. A weft knit is not run-resistant, but stretches more. This is especially true if spools of spandex are processed from separate spool containers and interwoven through the cylinder with cotton yarn, giving the finished product more flexibility and making it less prone to having a 'baggy' appearance. The average t-shirt is a weft knit.

Finishing- processing of Textiles

The woven cotton fabric in its loom-state not only contains impurities, including warp size, but requires further treatment in order to develop its full textile potential. Furthermore, it may receive considerable added value by applying one or more finishing processes.

Desizing

Depending on the size that has been used, the cloth may be steeped in a dilute acid and then rinsed, or enzymes may be used to break down the size.

Scouring

Scouring, is a chemical washing process carried out on cotton fabric to remove natural wax and non-fibrous impurities (e.g. the remains of seed fragments) from the fibres and any added soiling or dirt. Scouring is usually carried in iron vessels called kiers. The fabric is boiled in an alkali, which forms a soap with free fatty acids (saponification). A kier is usually enclosed, so the solution of sodium hydroxide can be boiled under pressure, excluding oxygen which would degrade the cellulose in the fibre. If the appropriate reagents are used, scouring will also remove size from the fabric although desizing often precedes scouring and is considered to be a separate process known as fabric preparation. Preparation and scouring are prerequisites to most of the other finishing processes. At this stage even the most naturally white cotton fibres are yellowish, and bleaching, the next process, is required.

Bleaching

Bleaching improves whiteness by removing natural coloration and remaining trace impurities from the cotton; the degree of bleaching necessary is determined by the required whiteness and absorbency. Cotton being a vegetable fibre will be bleached using an oxidizing agent, such as dilute sodium hypochlorite or dilute hydrogen peroxide. If the fabric is to be dyed a deep shade, then lower levels of bleaching are acceptable, for example. However, for white bed sheetings and medical applications, the highest levels of whiteness and absorbency are essential.

Mercerising

A further possibility is mercerizing during which the fabric is treated with caustic soda solution to cause swelling of the fibres. This results in improved lustre, strength and dye affinity. Cotton is mercerized under tension, and all alkali must be washed out before the tension is released or shrinkage will take place. Mercerizing can take place directly on grey cloth, or after bleaching.

Many other chemical treatments may be applied to cotton fabrics to produce low flammability, crease resist and other special effects but four important non-chemical finishing treatments are:

Singeing

Singeing is designed to burn off the surface fibres from the fabric to produce smoothness. The fabric passes over brushes to raise the fibres, then passes over a plate heated by gas flames.

Raising

Another finishing process is raising. During raising, the fabric surface is treated with sharp teeth to lift the surface fibres, thereby imparting hairiness, softness and warmth, as in flannelette.

Calendering

Calendering is the third important mechanical process, in which the fabric is passed between heated rollers to generate smooth, polished or embossed effects depending on roller surface properties and relative speeds.

Shrinking (Sanforizing)

Finally, mechanical shrinking (sometimes referred to as sanforizing), whereby the fabric is forced to shrink width and/or lengthwise, creates a fabric in which any residual tendency to shrink after subsequent laundering is minimal.

Dyeing

Finally, cotton is an absorbent fibre which responds readily to colouration processes. Dyeing, for instance, is commonly carried out with an anionic direct dye by completely immersing the fabric (or yarn) in an aqueous dyebath according to a prescribed procedure. For improved fastness to washing, rubbing and light, other dyes such as vats and reactives are commonly used. These require more complex chemistry during processing and are thus more expensive to apply.

Printing

Printing, on the other hand, is the application of colour in the form of a paste or ink to the surface of a fabric, in a predetermined pattern. It may be considered as localised dyeing. Printing designs onto already dyed fabric is also possible.

Economic, Environmental and Political Consequences of Cotton Manufacture

The growth of cotton is divided into two segments i.e. organic and genetically modified. Cotton crop provides livelihood to millions of people but its production is becoming expensive because of high water consumption, use of expensive pesticides, insecticides and fertiliser. Genetically Modified products aim to increase disease resistance and reduce the water required. The organic sector was worth $583 million. Genetically Modified cotton, in 2007, occupied 43% of cotton growing areas.

Cotton is farmed intensively and uses large amounts of fertilizer and 25% of the world's insecticides. Native Indian varieties of cotton were rainwater fed, but modern hybrids used for the mills need irrigation, which spreads pests. The 5% of cotton-bearing land in India uses 55% of all pesticides used in India.

The consumption of energy in form of water and electricity is relatively high, especially in processes like washing, de-sizing, bleaching, rinsing, dyeing, printing, coating and finishing. Processing is time consuming. The major portion of water in textile industry is used for wet processing of textile (70 per cent). Approximately 25 per cent of energy in the total textile production like fibre production, spinning, twisting, weaving, knitting, clothing manufacturing etc. is used in dyeing. About 34 per cent of energy is consumed in spinning, 23 per cent in weaving, 38 per cent in chemical wet processing and five per cent in miscellaneous processes. Power dominates consumption pattern in spinning and weaving, while thermal energy is the major factor for chemical wet processing.

Before mechanisation, cotton was harvested manually by farmers in India and by African slaves in America. In 2012 Uzbekistan was a major exporter of cotton and uses manual labour during the harvest. Human rights groups claim that health care professionals and children are forced to pick cotton.

Processing of Other Vegetable Fibres

Flax

Flax is a bast fibre, which means it comes in bundles under the bark of the Linum usitatissimum plant. The plant flowers and is harvested.

- Retting
- Breaking
- Scutching
- Hackling or combing

It is now treated like cotton.

Jute

Jute is a bast fibre, which comes from the inner bark of the plants of the Corchorus genus. It is retted like flax, sundried and baled. When spinning a small amount of oil must be added to the fibre. It can be bleached and dyed. It was used for sacks and bags but is now used for the backing for carpets. Jute can be blended with other fibres to make composite fabrics and work continues in Bangladesh to refine the processes and extend the range of usage possible. In the 1970s, jute-cotton composite fabrics were known as *jutton* fabrics.

Hemp

Hemp is a bast fibre from the inner bark of Cannabis sativa. It is difficult to bleach, it is used for making cord and rope.

- Retting
- Separating
- Pounding

Other Bast Fibres

These bast fibres can also be used: kenaf, urena, ramie, nettle.

Other Leaf Fibres

Sisal is the main leaf fibre used; others are: abacá and henequen.

Processing of Protein Fibres

Wool

Wool comes from domesticated sheep. It forms two products, woolens and worsteds. The sheep has two sorts of wool and it is the inner coat that is used. This can be mixed with wool that has been recovered from rags. Shoddy is the term for recovered wool that is not matted, while mungo comes from felted wool. Extract is recovered chemically from mixed cotton/wool fabrics.

The fleece is cut in one piece from the sheep.This is then skirted to remove the soiled wool, and baled. It is graded into long wool where the fibres can be up to 15 in, but anything over 2.5 inches is suitable for combing into worsteds. Fibres less than that form short wool and are described as clothing or carding wool.

At the mill the wool is scoured in a detergent to remove grease (the yolk) and impurities. This is done mechanically in the opening machine. Vegetable matter can be removed chemically using sulphuric acid (carbonising). Washing uses a solution of soap and sodium carbonate. The wool is oiled before carding or combing.

- Woollens: Use noils from the worsted combs, mungo and shoddy and new short wool

- Worsteds

Combing: Oiled slivers are wound into laps, and placed in the circular comber. The worsted yarn gathers together to form a top. The shorter fibres or noils remain behind and are removed with a knife.

- Angora

Silk

The processes in silk production are similar to those of cotton but take account that reeled silk is a continuous fibre. The terms used are different.

- Opening bales. Assorting skeins: where silk is sorted by colour, size and quality, scouring: where the silk is washed in water of 40 degrees for 12 hours to remove the natural gum, drying: either by steam heating or centrifuge, softening: by rubbing to remove any remaining hard spots.

- Silk throwing (winding). The skeins are placed on a reel in a frame with many others. The silk is wound onto spools or bobbins.

  - Doubling and twisting. The silk is far too fine to be woven, so now it is doubled and twisted to make the warp, known as organzine and the weft, known as tram. In organzine each single is given a few twists per inch (tpi), and combine with several other singles counter twisted hard at 10 to 14 tpi. In tram the two singles are doubled with each other with a light twist, 3 to 6 tpi. Sewing thread is two tram threads, hard twisted, and machine-twist is made of three hard-twisted tram threads. Tram for the crepe process is twisted at up to 80 tpi to make it 'kick up'.

  - Stretching. The thread is tested for consistent size. Any uneven thickness is stretched out. The resulting thread is reeled into containing 500 yd to 2500 yd. The skeins are about 50 in in loop length.

- Dyeing: the skeins are scoured again, and discoloration removed with a sulphur process. This weakens the silk. The skeins are now tinted or dyed. They are dried and rewound onto bobbins, spools and skeins. Looming, and the weaving process on power looms is the same as with cotton.

- Weaving. The organzine is now warped. This is a similar process to in cotton. Firstly, thirty threads or so are wound onto a warping reel, and then using the warping reels, the threads are beamed. A thick layer of paper is laid between each layer on the beam to stop entangling.

Processing of Synthetic Fibres

Discussion of Types of Synthetic Fibers

Synthetic fibers are the result of extensive development by scientists to improve upon the naturally occurring animal and plant fibers. In general, synthetic fibers are created by forcing, or extruding, fiber forming materials through holes (called spinnerets) into the air, thus forming a thread. Before synthetic fibers were developed, cellulose fibers were made from natural cellulose, which comes from plants.

The first artificial fiber, known as art silk from 1799 onwards, became known as viscose around 1894, and finally rayon in 1924. A similar product known as cellulose acetate was discovered in 1865. Rayon and acetate are both artificial fibers, but not truly synthetic, being made from wood. Although these artificial fibers were discovered in the mid-nineteenth century, successful modern manufacture began much later in the 1930s. Nylon, the first synthetic fiber, made its debut in the United States as a replacement for silk, and was used for parachutes and other military uses.

The techniques used to process these fibers in yarn are essentially the same as with natural fibers, modifications have to be made as these fibers are of great length, and have no texture such as the scales in cotton and wool that aid meshing.

Textile Manufacturing by Pre-industrial Methods

Textile manufacturing is one of the oldest human activities. The oldest known textiles date back to about 5000 B.C. In order to make textiles, the first requirement is a source of fibre from which a yarn can be made, primarily by spinning. The yarn is processed by knitting or weaving to create cloth. The machine used for weaving is the loom. Cloth is finished by what are described as wet processes to become fabric. The fabric may be dyed, printed or decorated by embroidering with coloured yarns.

The three main types of fibres are natural vegetable fibres, animal protein fibres and artificial fibres. Natural vegetable fibres include cotton, linen, jute and hemp. Animal protein fibres include wool and silk. Man-made fibres (made by industrial processes) including nylon, polyester will be used in some hobbies and hand crafts and in the developed world.

Almost all commercial textiles are produced by industrial methods. Textiles are still produced by pre-industrial processes in village communities in Asia, Africa and South America. Creating textiles using traditional manual techniques is an artisan craft practised as a hobby in Europe and North America.

Yarn Formation

Vegetable Fibres

Flax

The preparations for spinning is similar across most plant fibres, including Flax and Hemp. Flax is the fibre used to create linen. Cotton is handled differently since it uses the fruit of the plant and not the stem.

Harvesting

Flax is pulled out of the ground about a month after the initial blooming when the lower part of the plant begins to turn yellow, and when the most forward of the seeds are found in a soft state. It is pulled in handfuls and several handfuls are tied together with slip knot into a 'beet'. The string is tightened as the stalks dry. The seed heads are removed and the seeds collected, by threshing and winnowing.

Retting

Retting is the process of rotting away the inner stalk, leaving the outer fibres intact. A standing pool of warm water is needed, into which the beets are submerged. An acid is produced when retting, and it would corrode a metal container.

At 80 °F (27 °C), the retting process takes 4 or 5 days, it takes longer when colder. When the retting is complete the bundles feel soft and slimy, The process can be overdone, and the fibres rot too.

Dressing the Flax

Dressing is removing the fibres from the straw and cleaning it enough to be spun. The flax is broken, scutched and hackled in this step.

Breaking The process of breaking breaks up the straw into short segments. The beets are untied and fed between the beater of the breaking machine , the set of wooden blades that mesh together when the upper jaw is lowered.

Breaking flax in pre-revolutionary Perm, Russia

Scutching In order to remove some of the straw from the fibre a wooden scutching knife is scaped down the fibres while they hang vertically.

Heckling Fibre is pulled through various sized heckling combs. A Heckling comb is a bed of sharp, long-tapered, tempered, polished steel pins driven into wooden blocks at regular spacing. A good progression is from 4 pins per square inch, to 12, to 25 to 48 to 80. The first three will remove the straw, and the last two will split and polish the fibres. Some of the finer stuff that comes off in the last heckles can be carded like wool and spun. It will produce a coarser yarn than the fibres pulled through the heckles because it will still contain some straw.

Spinning

Flax being spun from a distaff

Flax can either be spun from a distaff, or from the spinner's lap. Spinners keep their fingers wet when spinning, to prevent forming fuzzy thread. Usually singles are spun with an "S" twist. After flax is spun it is washed in a pot of boiling water for a couple of hours to set the twist and reduce fuzziness.

Many handspinners, will buy a roving of flax. This roving is spun in the same manner as above. The rovings may come with very long fibres (4 to 8 inches), or much shorter fibres (2 to 3 inches).

Cotton

Picking cotton in Oklahoma, USA, in the 1890s

Cotton is a soft, fluffy staple fibre that grows in a boll, or protective capsule, around the seeds of cotton plants of the genus *Gossypium*. The fibre is almost pure cellulose.

The plant is a shrub native to tropical and subtropical regions around the world, including the Americas, Africa, and India. The greatest diversity of wild cotton species is found in Mexico, followed by Australia and Africa. Cotton was independently domesticated in the Old and New Worlds. The most favoured cottons are the ones with the longest staple as they can be spun into the finest thread. Sea Island and Egyptian are two of these. Surat an Indian species has a short staple. Hand operated methods of processing remained the preferred way of spinning and weaving the very finest threads and fabrics into the third quarter of the nineteenth century.

Yucca

Yucca fibres were at one time widely used throughout Central America for many things. Currently they are mainly used to make twine.Yucca leaves are harvested and then cut to a standard size. The leaves are crushed in between two large rollers producing the fibres which are bundled up and dried in the sun over trellises. The dried fibres are combined into rolags. At this point it is ready to spin. The waste, a pulpy liquid that stinks, can be used as a fertilizer.

Animal Protein Fibres

Wool

Wool is a protein based fibre, being the coat of a sheep. The wool is removed by shearing.

Sheep Shearing

A half sheared sheep.

Shearing can be done with use of hand-shears or powered shears. Professional sheep shearers can shear a sheep in under a minute, without nicking the sheep.

The fleece is removed in one piece. Second cuts can be made but produce only short fibres, which are more difficult to spin. Primitive breeds, like the Scottish Soay sheep have to be plucked, not sheared, as the kemps are still longer than the soft fleece, (a process called rooing).

Skirting

Skirting is disposing of all wool that is unsuitable for spinning. Recovering can be attempted. It can also be done at the same time as carding.

1595 painting illustrating Leiden textile workers

Cleaning

The wool is cleaned. At this point the fleece is full of lanolin and often contains extraneous vegetable matter, such as sticks, twigs, burrs and straw. These may all be removed, though lanolin may be left in the wool till after the spinning, a technique known as spinning 'in the grease'. Indeed if the fabric is to be water repellent, lanolin is not removed at any stage.

Washing the wool at this stage can be a tedious process. Some people wash it a small handful at a time very carefully, and then set it out to dry on a table in the sun. Others will wash the whole fleece. Lanolin is removed by soaking the fleece in very hot water. If the fleece gets agitated, it will become felt, and then spinning is impossible. Felting, when done on purpose (with needles, chemicals, or simply rubbing the fibres against each other), can be used to create garments.

Carding or Combing

It is possible to spin directly from a clean fleece, but it is much easier to spin a carded fleece. Carding by hand yields a rolag, a loose woollen roll of fibres. Using a drum carder yields a bat, which is a mat of fibres in a flat, rectangular shape. Carding mills return the fleece in a roving, which is a stretched bat; it is very long and often the thickness of a wrist.A pencil roving is a roving thinned to the width of a pencil. It can used for knitting without any spinning, or for apprentices.

Combing is another method to align the fibres parallel to the yarn, and thus is good for spinning a worsted yarn, whereas the rolag from handcards produces a woolen yarn.

Spinning

Hand spinning can be done by using a spindle or the spinning wheel. Spinning turns the carded wool fibres into yarn which can then be directly woven, knitted (flat or circular), crocheted, or by other means turned into fabric or a garment.

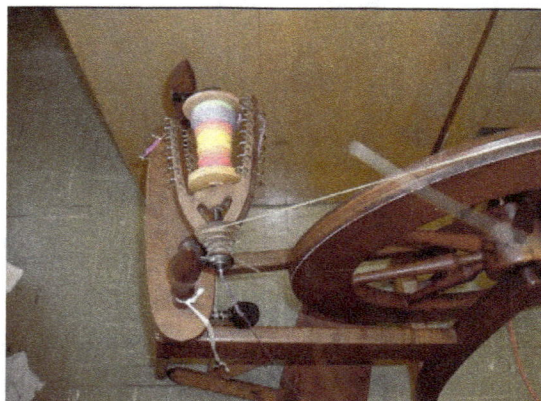

A spinning wheel used to make yarn.

Spindle and distaff

The spinning wheel collects the yarn on a bobbin. A woollen yarn is lightly spun so it is airey, and is a good insulator and suitable for knitting, while a worsted yarn is spun tight to exclude air, and has greater strength and is suited to weaving..

A niddy noddy ready to have a skein wound on it.

Once the bobbin is full, the hobby spinner either puts on a new bobbin, or forms a skein, or balls the yarn. A skein is a coil of yarn twisted into a loose knot. Yarn is skeined using a niddy-noddy or other type of skein -winder. Yarn is rarely balled directly after spinning, it will be stored in skein form, and transferred to a ball only if needed. Knitting from a skein, is difficult as the yarn forms knots, in this case it is best to ball. Yarn to be plied is left on the bobbin.

S and Z twists

A lazy kate with bobbins on it in preparation for plying.

A skein is either formed by the hobby spinner, on a niddy noddy or some other type of skein winder. Traditionally niddy-noddys looked like an uppercase "i", with the bottom half rotated 90 degrees. Hobby spinning wheel manufactures also make niddy-noddys that attach onto the spinning wheel for faster skein winding.

Plying

Plying yarn is when one takes a strand of spun yarn (one strand is often called a single) and spins it together with other strands in order to make a thicker yarn.

Regular plying consists of taking two or more singles and twisting them together, the against their twist. This can be done on a spinning wheel or on a spindle. If the yarn was spun clockwise (which is called a "Z" twist), to ply, the wheel must spin counter-clockwise (an "S" twist). This is the most common way. When plying from bobbins a device called a lazy kate is often used to hold them.

Most hobby spinners (who use spinning wheels) ply from bobbins. This is easier than plying from balls because there is less chance for the yarn to become tangled and knotted if it is simply unwound from the bobbins. So that the bobbins can unwind freely, they are put in a device called a lazy kate, or sometimes simply *kate*. The simplest lazy kate consists of wooden bars with a metal rod running between them. Most hold between three and four bobbins. The bobbin sits on the metal rod. Other lazy kates are built with devices that create an adjustable amount of tension, so that if the yarn is jerked, a whole bunch of yarn is not wound off, then wound up again in the opposite direction. Some spinning wheels come with a built in lazy kate.

Navajo plying consists of making large loops, similar to crocheting.A loop about 8 inches long is made on the leader the end on the leader. (A leader is the string left on the bobbin to spin off.) The three strands together are spun in the opposite direction. When a third of the loop remains, a new loop is created and the spinning continues. The process is repeated until the yarn is all plied. The advantage of this method is that only one single is needed and if the single is already dyed this technique allows it to be plied without ruining the color scheme. This technique also allows the spinner to try to match up thick and thin spots in the yarn, thus making for a smoother end product.

Washing

If the lanolin is unwanted, and has not already been washed out, this is done now. The skein is tied in six points and steeped overnight in detergent, it is rinsed and air-dried, and re-skeined.

Washing the skins and grading the wool, painting of the wool trade in Leiden, c. 1595

unless the lanolin is to be left in the cloth as a water repellent. When washing a skein it works well to let the wool soak in soapy water overnight, and rinse the soap out in the morning. Dishwashing detergents are commonly used, and a special laundry detergent designed for washing wool is not required. The dishwashing detergent works and does not harm the wool. After washing, let the wool dry (air drying works best). Once it is dry, or just a bit damp, one can stretch it out a bit on a niddy-noddy. Putting the wool back on the niddy-noddy makes for a nicer looking finished skein. Before taking a skein and washing it, the skein must be tied up loosely in about six places. If the skein is not tied up, it will be very hard to unravel when done washing.

Silk

Silk fabric was first developed in ancient China, with some of the earliest examples found as early as 3500 BC.

Cultivation

Silk moths lay eggs on specially prepared paper. The eggs hatch and the caterpillars (silkworms) are fed on fresh mulberry leaves. After about 35 days and 4 moltings, the caterpillars are 10,000 times heavier than when hatched and are ready to begin spinning a cocoon. A straw frame is placed over the tray of caterpillars, and each caterpillar begins spinning a cocoon by moving its head in a pattern. Two glands produce liquid silk and force it through openings in the head called spinnerets. Liquid silk is coated in sericin, a water-soluble protective gum, and solidifies on contact with the air. Within 2–3 days, the caterpillar spins about 1 mile of filament and is completely encased in a cocoon.

Harvesting

The silk farmers then kill most caterpillars by heat, leaving some to metamorphose into moths to breed the next generation of caterpillars. Harvested cocoons are then soaked in boiling water to soften the sericin holding the silk fibres together in a cocoon shape.

Throwing

The fibres are then unwound to produce a continuous thread. Since a single thread is too fine and fragile for commercial use, anywhere from three to ten strands are spun together to form a single thread of silk. Colloquially silk throwing can be used to refer to the whole process: reeling, throwing and doubling, and silk throwsters would speak of throwing as twisting or spinning.

1843 Illustration

Silk throwing was originally a hand process relying on a turning a wheel (the gate) that twisted four threads while a helper who would be a child, ran the length of a shade, hooked the threads on stationary pins (the cross)and ran back to start the process again. The shade would be a between 23 and 32m long. The process was described in detail to Lord Shaftesbury's Royal Commission of Inquiry into the Employment of Children in 1841:

For twisting it is necessary to have what are designated shades which are buildings of at least 30 or 35 yards in length, ... the upper storey is generally occupied by children, ... or grown women as 'piecers', 'winders' and 'doublers' attending to their reels and bobbins [which is], driven by the exertions of one man... He (the boy) takes first a rod containing four bobbins of silk from the twister who stands at his gate or wheel, and having fastened the ends, runs to the 'cross' at the extreme end of the room, round which he passes the threads of each bobbin and returns to the 'gate'. He is despatched on a second expedition of the same kind, and returns as before, he then runs up to the cross and detaches the threads and comes to the roller. Supposing the master to make twelve rolls a day, the boy necessarily runs fourteen miles, and this is barefooted.

In 1700, the Italians were the most technologically advanced throwsters in Europe and had developed two machines capable of winding the silk onto bobbins while putting a twist in the thread. They called the throwing machine, a *filatoio*, and called the doubler, a *torcitoio*. There is an illustration of a circular hand-powered throwing machine drawn in 1487 with 32 spindles. The first evidence of an externally powered filatoio comes from the thirteenth century, and the earliest illustration from around 1500. Filatorios and torcitoios contained parallel circular frames that revolved round each other on a central axis. The speed of the relative rotation determined the twist. Silk would only cooperate in the process if the temperature and humidity were high.

Fabric Formation

Once the fibre has been turned into yarn the process of making cloth is much the same for any type of fibre, be it animal or plant.

Weaving

The earliest weaving was done without a loom.

Loom

Elements of a foot-treadle floor loom

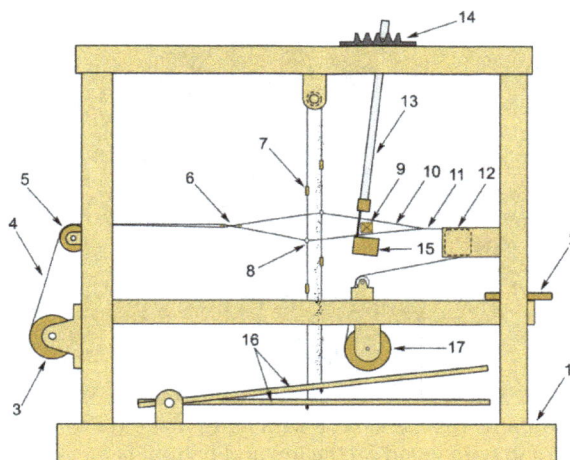

1. Wood frame
2. Seat for weaver
3. Warp beam
4. Warp threads
5. Back beam or platen
6. Rods – used to make a shed
7. Heddle bar
8. heddle- the eye
9. shuttle with weft yarn
10. Shed
11. Completed fabric
12. Breast beam
13. Batten with reed comb
14. Batten adjustment

15. Lathe

16. Treadles

17. Cloth roll

In general the supporting structure of the loom is called the *frame*. It provides the means of fixing the length-wise threads, called the warp, and keeping them under tension. The warp threads are wound on a roller called the *warp beam*, and attached to the *cloth beam* which will hold the finished material. Because of the tension the warp threads are under, they need to be strong.

A picture taken from the back of a loom. The metal rods with holes that have the yarn running through them are the heddles. Further back, the metal comb with wood on the top and bottom is the reed. The shed is the gap between the two sets of yarn.

The thread that is woven through the warp is called the weft. The weft is threaded through the warp using a shuttle. The original *hand-loom* was limited in width by the weaver's reach, because of the need to throw the shuttle from hand to hand. The invention of the flying shuttle with its *fly cord* and *picking sticks* enabled the weaver to pass the shuttle from a *box* at either side of the loom with one hand, and across a greater width. The invention of the drop box allowed a weaver to use multiple shuttles to carry different wefts.

Alternating sets of threads are lifted by connecting them with string or wires called heddles to another bar, called the *shaft* (or *heddle bar* or *heald*). Heddles, shafts and the *couper* (lever to lift the assembly) are called the *harness* — the harness provides for mechanical operation using foot- or hand-operated *treadles*. After passing a weft thread through the warp, a reed comb is used to *beat* (compact) the woven weft.

To prepare to weave, the warp must be made. By hand this is done with the help of a warping board. The length the warp is made is about a quarter to half yard more than the amount of cloth needed. Warping boards come in a variety of shapes, from the two nearest door handles to a board with pegs on it, or a device called a warping mill that looks similar to a swift. Warping the loom, mean threading each *end* through an eye in a heddle, and then *sleying it through the reed*. The warp is *set* (verb) at X ends per inch. It then has a *sett* (noun) of X ends per inch. The weft is measured in picks per inch.

Knitting

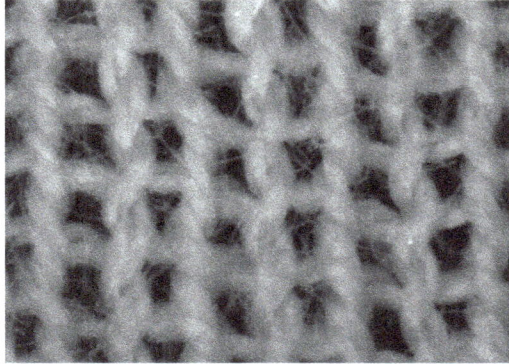

The front side of a plainly knitted object

Knitting needles

Hand knitting can either be done "flat" or "in the round". Flat knitting is done on a set of single point knitting needles, and the knitter goes back and forth, adding rows. In Circular knitting, or "knitting in the round", the knitter knits around a circle, creating a tube. This can be done with a set of four double pointed needles or a single circular needle.

A knitted object will unravel easily if the top has not been secured. Knitted objects also stretch easily in all directions, whereas woven fabric only stretches on the bias.

Crocheting

Irish crocheted lace

Crocheting differs largely from knitting in that there is only one loop, not the multitude as knitting has. Also, instead of knitting needles, a crochet hook is used.

Lace Making

A lace fabric is lightweight openwork fabric, patterned, with open holes in the work. The holes can be formed via removal of threads or cloth from a previously woven fabric, but more often lace is built up from a single thread and the open spaces are created as part of the lace fabric. Lace may be crocheted, or knitted.

Textile Finishing

Wet Processes

Embroidery

Embroidery – threads which are added to the surface of a finished textile.

Embroidery is the handicraft of decorating fabric or other materials with needle and thread or yarn. Embroidery may also incorporate other materials such as metal strips, pearls, beads, quills, and sequins. Embroidery is most often used on caps, hats, coats, blankets, dress shirts, denim, stockings, and golf shirts. Embroidery is available with a wide variety of thread or yarn color.

References

- Fowler, Alan (2003), Lancashire Cotton Operatives and Work, 1900-1950: A Social History of Lancashire Cotton Operatives in the Twentieth Century, Ashgate Publishing, Ltd., p. 90, ISBN 0-7546-0116-1.
- Calladine, Anthony; Fricker (1993). East Cheshire Textile Mills. London: Royal Commission on Historical Monuments of England. ISBN 1-873592-13-2.

2

Processing of Textile

This chapter focuses on the various processes employed on cotton fabric in its loom state to remove impurities and to improve its quality. The reader is introduced to processes like desizing, scouring, bleaching, mercerizing, singeing, raising, calendering, sanforizing, dyeing and textile printing. The chapter studies these multiple facets of textile processing in-depth and provides the reader with comprehensive information.

Desizing

Desizing is the process of removing the size material from the warp yarns after the textile fabric is woven.

Sizing Agents

Sizing agents are selected on the basis of type of fabric, environmental friendliness, ease of removal, cost considerations, effluent treatment, etc.

Natural Sizing Agents

Natural sizing agents are based on natural substances and their derivatives:

- Starch and starch derivatives; native starch, degradation starch and chemically modified starch products
- Cellulosic derivatives; carboxymethylcellulose (CMC), methylcellulose and oxyethylcellulose
- Protein-based starches; glue, gelatin, albumen

Synthetic Sizing Agents

- Polyacrylates,
- Modified polyesters,
- Polyvinyl alcohols (PVA),
- Styrene/maleic acid copolymers.

Desizing Processes

Desizing, irrespective of what the desizing agent is, involves impregnation of the fabric with the desizing agent, allowing the desizing agent to degrade or solubilise the size material, and finally to wash out the degradation products. The major desizing processes are:

- Enzymatic desizing of starches on cotton fabrics

- Oxidative desizing

- Acid desizing

- Removal of water-soluble sizes

Enzymatic Desizing

Enzymatic desizing is the classical desizing process of degrading starch size on cotton fabrics using enzymes. Enzymes are complex organic, soluble bio-catalysts, formed by living organisms, that catalyze chemical reaction in biological processes. Enzymes are quite specific in their action on a particular substance. A small quantity of enzyme is able to decompose a large quantity of the substance it acts upon. Enzymes are usually named by the kind of substance degraded in the reaction it catalyzes.

Amylases are the enzymes that hydrolyses and reduce the molecular weight of amylose and amylopectin molecules in starch, rendering it water-soluble enough to be washed off the fabric.

Effective enzymatic desizing require strict control of pH, temperature, water hardness, electrolyte addition and choice of surfactant.

Oxidative Desizing

In oxidative desizing, the risk of damage to the cellulose fiber is very high, and its use for desizing is increasingly rare. Oxidative desizing uses potassium or sodium persulfate or sodium bromite as an oxidizing agent.

Acid Desizing

Cold solutions of dilute sulphuric or hydrochloric acids are used to hydrolyze the starch, however, this has the disadvantage of also affecting the cellulose fiber in cotton fabrics.

Removal of Water-soluble Sizes

Fabrics containing water-soluble sizes can be desized by washing using hot water, perhaps containing wetting agents (surfactants) and a mild alkali. The water replaces the size on the outer surface of the fiber, and absorbs within the fiber to remove any fabric residue.

Dyeing

Dyeing is the process of adding color to textile products like fibers, yarns, and fabrics. Dyeing is normally done in a special solution containing dyes and particular chemical material. After dyeing, dye molecules have uncut chemical bond with fiber molecules. The temperature and time controlling are two key factors in dyeing. There are mainly two classes of dye, natural and man-made.

Pigments for sale at a market in Goa, India.

Cotton being dyed manually in contemporary India.

The primary source of dye, historically, has generally been nature, with the dyes being extracted from animals or plants. Since the mid-19th century, however, humans have produced artificial dyes to achieve a broader range of colors and to render the dyes more stable to resist washing and general use. Different classes of dyes are used for different types of fiber and at different stages of the textile production process, from loose fibers through yarn and cloth to complete garments.

Acrylic fibers are dyed with basic dyes, while nylon and protein fibers such as wool and silk are dyed with acid dyes, and polyester yarn is dyed with disperse dyes. Cotton is dyed with a range of dye types, including vat dyes, and modern synthetic reactive and direct dyes.

Etymology

The word dye is from Middle English *deie* and from Old English *dag* and *dah*. The first known use of the word dye was before the 12th century.

History

The earliest dyed flax fibers have been found in a prehistoric cave in the Republic of Georgia and date back to 34,000 BC. More evidence of textile dyeing dates back to the Neolithic period at the large Neolithic settlement at Çatalhöyük in southern Anatolia, where traces of red dyes, possibly from ocher, an iron oxide pigment derived from clay, were found. In China, dyeing with plants, barks, and insects has been traced back more than 5,000 years. Early evidence of dyeing comes from Sindh province in Pakistan, where a piece of cotton dyed with a vegetable dye was recovered from the archaeological site at Mohenjo-daro (3rd millennium BCE). The dye used in this case was madder, which, along with other dyes such as indigo, was introduced to other regions through trade. Natural insect dyes such as Cochineal and kermes and plant-based dyes such as woad, indigo and madder were important elements of the economies of Asia and Europe until the discovery of man-made synthetic dyes in the mid-19th century. The first synthetic dye was William Perkin's mauveine in 1856, derived from coal tar. Alizarin, the red dye present in madder, was the first natural pigment to be duplicated synthetically in 1869, a development which led to the collapse of the market for naturally grown madder. The development of new, strongly colored synthetic dyes followed quickly, and by the 1870s commercial dyeing with natural dyestuffs was disappearing.

Dyeing in Fes, Morocco.

Methods

Dyes are applied to textile goods by dyeing from dye solutions and by printing from dye pastes. Methods include direct application and yarn dyeing.

Direct Application

The term "direct dye application" stems from some dyestuff having to be either fermented as in the case of some natural dye or chemically reduced as in the case of synthetic vat and sulfur dyes before being applied. This renders the dye soluble so that it can be absorbed by the fiber since the insoluble dye has very little substantivity to the fiber. Direct dyes, a class of dyes largely for dyeing cotton, are water-soluble and can be applied directly to the fiber from an aqueous solution. Most other classes of synthetic dye, other than vat and surface dyes, are also applied in this way.

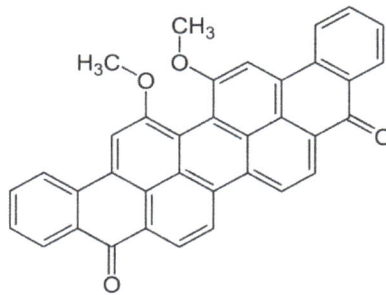

Chemical structure of Vat Green 1, a type of vat dye

The term may also be applied to dyeing without the use of mordants to fix the dye once it is applied. Mordants were often required to alter the hue and intensity of natural dyes and improve color fastness. Chromium salts were until recently extensively used in dying wool with synthetic mordant dyes. These were used for economical high color fastness dark shades such as black and navy. Environmental concerns have now restricted their use, and they have been replaced with reactive and metal complex dyes that do not require mordant.

Yarn dyeing

Dyed Wool Reels (CSIRO)

There are many forms of yarn dyeing. Common forms are the package form and the hanks form. Cotton yarns are mostly dyed at package form, and acrylic or wool yarn are dyed at hank form. In the continuous filament industry, polyester or polyamide yarns are always dyed at package form, while viscose rayon yarns are partly dyed at hank form because of technology.

The common dyeing process of cotton yarn with reactive dyes at package form is as follows:

1. The raw yarn is wound on a spring tube to achieve a package suitable for dye penetration.

2. These softened packages are loaded on a dyeing carrier's spindle one on another.

3. The packages are pressed up to a desired height to achieve suitable density of packing.

4. The carrier is loaded on the dyeing machine and the yarn is dyed.

5. After dyeing, the packages are unloaded from the carrier into a trolley.

6. Now the trolley is taken to hydro extractor where water is removed.

7. The packages are hydro extracted to remove the maximum amount of water leaving the desired color into raw yarn.

8. The packages are then dried to achieve the final dyed package.

After this process, the dyed yarn packages are packed and delivered.

History of Garment Dyeing

Garment dyeing is the process of dyeing fully fashioned garments subsequent to manufacturing, as opposed to the conventional method of manufacturing garments from pre-dyed fabrics.

Up until the mid-1970s the method was rarely used for commercial clothing production. It was used domestically, to overdye old, worn and faded clothes, and also by resellers of used or surplus military clothing. The first notable industrial use of the technique was made by Benetton, which garment dyed its Shetland wool knitwear.

Complex Garment Dyeing

In the mid-1970s the Bologna clothing designer Massimo Osti began experimenting with the garment dyeing technique. His experimentation over the next decade, led to the pioneering of not just the industrial use of "traditional" garment dyeing (dyeing simple cotton or wool garments) but, more importantly, the technique of "complex garment dyeing" which involved dyeing fully fashioned garments which had been constructed from multiple fabric or fiber types (e.g. a jacket made from both nylon and cotton, or linen, nylon and polyurethane coated cotton) in the same bath.

Up until its development by Massimo Osti (for his clothing brand C.P. Company), this technique had never been successfully industrially applied in any context. The complexity lay in developing both a practical and chemical understanding of how each fabric responded differently to the dye, how much it would shrink, how much colour it would absorb, developing entirely new forms of quality control to verify possible defects in fabric before dyeing etc.

Beyond the industrial advantages of the technique (purchasing fabric in one colour, white or natural, you may produce as many colours as you wish etc.), the artistic advantages of the technique were considerable and in many ways paved the way for the creation of the clothing style today known as Italian Sportswear. These advantages included

- the way in which different fibres absorbed the dye's colour allowed for the creation of incredibly nuanced differences in colour tones and a harmony that is impossible to achieve any other way

- the garment dyeing process naturally gave the fabric a "worn-in" hand allowing for the development of the casual and relaxed version of the classic menswear look which characterizes Italian sportswear

- the fact that each fabric and fibre type responds differently to the dye also produces a "deconstructed" effect, whereby the consumer's attention is drawn to the construction techniques of the jacket. For example: a more densely woven fabric absorbs the colour less intensely than a more open weave, the polyester stitching used for a cotton garment does not absorb any dye colour, producing a contrast colour stitch etc.

The disadvantages included:

- a relatively high failure rate for garments (between 5-10%)

- the difficulty in achieving a very tailored look due to difficulties in precisely calculating shrinkage rates

- high research and prototyping costs in order to understand how fabrics will behave in production

Today, whilst garment dyeing is a diffusely employed as an industrial technique around the globe, predominantly in the production of vintage style cotton garments and by fast fashion suppliers, complex garment dyeing is still practiced almost exclusively in Italy, by a handful of premium brands and suppliers who remain committed to the art.

Removal of Dyes

The dyer of the fountain "Weberbrunnen" in Monschau (Germany).

If things go wrong in the dyeing process, the dyer may be forced to remove the dye already applied by a process called "stripping" or discharging. This normally means destroying the dye with powerful reducing agents such as sodium hydrosulfite or oxidizing agents such as hydrogen peroxide or sodium hypochlorite. The process often risks damaging the substrate (fiber). Where possible, it is often less risky to dye the material a darker shade, with black often being the easiest or last option.

Textile Bleaching

JumTextile bleaching is one of the stages in the manufacture of textiles. All raw textile materials, when they are in natural form, are known as 'greige' material (pronounced grey-sh). This greige material will have its natural color, odor and impurities that are not suitable for clothing materials. Not only the natural impurities will remain on the greige material but also the add-ons that were made during its cultivation, growth and manufacture in the form of pesticides, fungicides, worm killers, sizes, lubricants, etc. The removal of these natural coloring matters and add-ons during the previous state of manufacturing is called scouring and bleaching.

Scouring

Scouring is the first process carried out with or without chemicals, at room temperature or at suitable higher temperatures with the addition of suitable wetting agents, alkali and so on. Scouring removes all the waxes, pectins and makes the textile material hydrophilic or water absorbent.

Bleaching

The next process of decolorization of greige material into a suitable material for next processing is called bleaching. Bleaching of textiles can be classified into oxidative bleaching and reductive bleaching.

Oxidative Bleaching

Generally oxidative bleachings are carried out using sodium hypochlorite, sodium chlorite or sulfuric acid. Natural fibres like cotton, ramie, jute, wool, bamboo are all generally bleached with oxidative methods.

Reductive Bleaching

Reductive bleaching is done with sodium hydrosulphite, a powerful reducing agent. Fibres like polyamides, polyacrylics and polyacetates can be bleached using reductive bleaching technology.

Optical Whiteners

After scouring and bleaching, optical brightening agents (OBAs) are applied to make the textile material appear a more brilliant white. These OBAs are available in different tints such as blue, violet and red.

Textile Printing

Textile printing is the process of applying colour to fabric in definite patterns or designs. In properly printed fabrics the colour is bonded with the fibre, so as to resist washing and friction. Textile printing is related to dyeing but in dyeing properly the whole fabric is uniformly covered with one colour, whereas in printing one or more colours are applied to it in certain parts only, and in sharply defined patterns.

In printing, wooden blocks, stencils, engraved plates, rollers, or silkscreens can be used to place colours on the fabric. Colourants used in printing contain dyes thickened to prevent the colour from spreading by capillary attraction beyond the limits of the pattern or design.

Design for a hand woodblock printed textile, showing the complexity of the blocks used to make repeating patterns. *Evenlode* by William Morris, 1883.

Evenlode block-printed fabric.

History

Woodblock printing is a technique for printing text, images or patterns used widely throughout East Asia and probably originating in China in antiquity as a method of printing on textiles and later paper. As a method of printing on cloth, the earliest surviving examples from China date to before 220.

Textile printing was known in Europe, via the Islamic world, from about the 12th century, and widely used. However, the European dyes tended to liquify, which restricted the use of printed patterns. Fairly large and ambitious designs were printed for decorative purposes such as wall-hangings and lectern-cloths, where this was less of a problem as they did not need washing. When paper became common, the technology was rapidly used on that for woodcut prints. Superior cloth was also imported from Islamic countries, but this was much more expensive.

The Incas of Peru, Chile and the Aztecs of Mexico also practiced textile printing previous to the Spanish Invasion in 1519; but owing to the lack of records before that date, it is impossible to say whether they discovered the art for themselves, or, in some way, learned its principles from the Asiatics.

During the later half of the 17th century the French brought directly by sea, from their colonies on the east coast of India, samples of Indian blue and white resist prints, and along with them, particulars of the processes by which they had been produced, which produced washable fabrics.

As early as the 1630s, the East India Company was bringing in printed and plain cotton for the English market. By the 1660s British printers and dyers were making their own printed cotton to sell at home, printing single colors on plain backgrounds; less colourful than the imported prints, but more to the taste of the British. Designs were also sent to India for their craftspeople to copy for export back to England. There were many dyehouses in England in the latter half of the 17th century, Lancaster being one area and on the River Lea near London another. Plain cloth was put through a prolonged bleaching process which prepared the material to receive and hold applied color; this process vastly improved the color durability of English calicoes and required a great deal of water from nearby rivers. One dyehouse was started by John Meakins, a London Quaker who lived in Cripplegate. When he died, he passed his dyehouse to his son-in-law Benjamin Ollive, Citizen and Dyer, who moved the dye-works to Bromley Hall where it remained in the family until 1823, known as Benjamin Ollive and Company, Ollive & Talwin, Joseph Talwin & Company and later Talwin & Foster. Samples of their fabrics and designs can be found in the Victoria and Albert Museum in London and the Smithsonian Copper-Hewett in New York.

On the continent of Europe the commercial importance of calico printing seems to have been almost immediately recognized, and in consequence it spread and developed there much more rapidly than in England, where it was neglected for nearly ninety years after its introduction. During the last two decades of the 17th century and the earlier ones of the 18th new dye works were started in France, Germany, Switzerland and Austria. It was only in 1738 that calico printing was first, practiced in Scotland, and not until twenty-six years later that Messrs Clayton of Bamber Bridge, near Preston, established in 1764 the first print-works in Lancashire, and thus laid the foundation of the industry.

From an artistic point of view most of the pioneer work in calico printing was done by the French. From the early days of the industry down to the latter half of the 20th century, the productions of the French printers in Jouy, Beauvais, Rouen, and in Alsace-Lorraine, were looked upon as representing all that was best in artistic calico printing.

Methods

Traditional textile printing techniques may be broadly categorised into four styles:

- Direct printing, in which colorants containing dyes, thickeners, and the mordants or substances necessary for fixing the colour on the cloth are printed in the desired pattern.

- The printing of a mordant in the desired pattern prior to dyeing cloth; the color adheres only where the mordant was printed.

- Resist dyeing, in which a wax or other substance is printed onto fabric which is subsequently dyed. The waxed areas do not accept the dye, leaving uncoloured patterns against a coloured ground.

- Discharge printing, in which a bleaching agent is printed onto previously dyed fabrics to remove some or all of the colour.

Resist and discharge techniques were particularly fashionable in the 19th century, as were combination techniques in which indigo resist was used to create blue backgrounds prior to block-printing of other colours. Modern industrial printing mainly uses direct printing techniques.

The printing process does involve several stages in order to prepare the fabric and printing paste, and to fix the impression permanently on the fabric:

- pre-treatment of fabric,

- preparation of colors,

- preparation of printing paste,

- impression of paste on fabric using printing methods,

- drying of fabric,

- fixing the printing with steam or hot air (for pigments),

- after process treatments.

Preparation of Cloth for Printing

Cloth is prepared by washing and bleaching. For a coloured ground it is then dyed. The cloth has always to be brushed, to free it from loose nap, flocks and dust that it picks up whilst stored. Frequently, too, it has to be sheared by being passed over rapidly revolving knives arranged spirally round an axle, which rapidly and effectually cuts off all filaments and knots, leaving the cloth perfectly smooth and clean and in a condition fit to receive impressions of the most delicate engraving. Some fabrics require very careful stretching and straightening on a stenter before they are wound around hollow wooden or iron centers into rolls of convenient size for mounting on the printing machines.

Preparation of Colours

The art of making colours for textile printing demands both chemical knowledge and extensive technical experience, for their ingredients must not only be in proper proportion to each other, but also specially chosen and compounded for the particular style of work in hand. A colour must comply to conditions such as shade, quality and fastness; where more colours are associated in the same design each must be capable of withstanding the various operations necessary for the development and fixation of the others. All printing pastes whether containing colouring matter or not are known technically as colours.

Colours vary considerably in composition. Most of them contain all the elements necessary for direct production and fixation. Some, however, contain the colouring matter alone and require various after-treatments; and others again are simply thickened mordants. A mordant is a metallic salt or other substance that combines with the dye to form an insoluble colour, either directly by steaming, or indirectly by dyeing. All printing colours require thickening to enable them to be transferred from colour-box to cloth without running or spreading beyond the limits of the pattern.

Thickening Agents

The printing thickeners used depend on the printing technique, the fabric and the particular dye-stuff . Typical thickening agents are starch derivatives, flour, gum arabic, guar gum derivatives, tamarind, sodium alginate, sodium polyacrylate, gum Senegal and gum tragacanth, British gum or dextrine and albumen.

Hot-water-soluble thickening agents such as native starch are made into pastes by boiling in double or jacketed pans. Most thickening agents used today are cold-soluble and require only extensive stirring.

Starch Paste

Starch paste is made from wheat starch, cold water, and olive oil, then thickened by boiling. Non-modified starch is applicable to all but strongly alkaline or strongly acid colours. With the former it thickens up to a stiff unworkable jelly. In the case of the latter, while mineral acids or acid salts convert it into dextrine, thus diminishing its viscosity or thickening power, organic acids do not have that effect. Today, modified carboxymethylated cold soluble starches are mainly used. These have a stable viscosity and are easy to rinse out of the fabric and give reproducible "short" paste rheology.

Flour paste is made in a similar way to starch paste; it is sometimes used to thicken aluminum and iron mordants. Starch paste resists of rice flour have been used for several centuries in Japan.

Gums

Gum arabic and gum Senegal are both traditional thickenings, but expense prevents them from being used for any but pale, delicate tints. They are especially useful thickenings for the light ground colours of soft muslins and sateens on account of the property they possess of dissolving completely out of the fibres of the cloth in the post-printing washing process, and they have a long flowing, viscous rheology, giving sharp print and good penetration in the cloth. Today guar gum and tamarind derivates offer a cheaper alternative.

British gum or dextrin is prepared by heating starch. It varies considerably in composition, sometimes being only slightly roasted and consequently only partly converted into dextrine, and at other times being highly torrefied, and almost completely soluble in cold water and very dark in colour. Its thickening power decreases and its gummy nature increases as the temperature at which it is roasted is raised. It is useful for strongly acid colours, and with the exception of gum Senegal, it is the best choice for strongly alkaline colours and discharges. Like the natural gums, it does not penetrate as well into the fibre of the cloth pr as deeply as pure starch or flour and is unsuitable for very dark, strong colours.

Gum tragacanth, or Dragon, which may be mixed in any proportion with starch or flour, is equally useful for pigment colours and mordant colours. When added to a starch paste it increases its penetrative power and adds to its softness without diminishing its thickness, making it easier to wash out of the fabric. It produces much more even colours than does starch paste alone. Used by itself it is suitable for printing all kinds of dark grounds on goods that are required to retain their soft clothy feel.

Starch always leaves the printed cloth somewhat harsh in feeling (unless modified carboxymethylated starches are used), but very dark colours can be obtained. Gum Senegal, gum arabic or modified guar gum thickening yield clearer and more even tints than does starch, suitable for lighter colours but less suited for very dark colours. (The gums apparently prevent the colours from combining fully with the fibers.) A printing stock solution is mostly a combination of modified starch and gum stock solutions.

Albumen

Albumen is both a thickening and a fixing agent for insoluble pigments. Chrome yellow, the ochres, vermilion and ultramarine are such pigments. Albumen is always dissolved in the cold, a process that takes several days when large quantities are required. Egg albumen is expensive and only used for the lightest shades. Blood albumen solution is used in cases when very dark colours are required to be absolutely fast to washing. After printing, albumen thickened colours are exposed to hot steam, which coagulates the albumen and effectually fixes the colours.

Printing Paste Preparation

Combinations of cold water-soluble carboxymethylated starch, guar gum and tamarind derivatives are most commonly used today in disperse screen printing on polyester. Alginates are used for cotton printing with reactive dyes, sodium polyacrylates for pigment printing, and in the case of vat dyes on cotton only carboxymethylated starch is used.

Formerly, colors were always prepared for printing by boiling the thickening agent, the colouring matter and solvents, together, then cooling and adding various fixing agents. At the present time, however, concentrated solutions of the colouring matters and other adjuncts are often simply added to the cold thickenings, of which large quantities are kept in stock.

Colours are reduced in shade by simply adding more stock (printing) paste. For example, a dark blue containing 4 oz. of methylene blue per gallon may readily be made into a pale shade by adding to it thirty times its bulk of starch paste or gum, as the case may be. The procedure is similar for other colours.

Before printing it is essential to strain or sieve all colours in order to free them from lumps, fine sand, and other impurities, which would inevitably damage the highly polished surface of the engraved rollers and result in bad printing. Every scratch on the surface of a roller prints a fine line on the cloth, and too much care, therefore, cannot be taken to remove, as far as possible, all grit and other hard particles from every colour.

Straining is usually done by squeezing the colour through filter cloths like artisanal fine cotton, silk or industrial woven nylon. Fine sieves can also be employed for colours that are used hot or are very strongly alkaline or acid.

Methods of Printing

There are seven distinct methods presently used to impress coloured patterns on cloth:

- Hand block printing

- Perrotine printing

- Engraved copperplate printing

- Roller, cylinder, or machine printing

- Stencil printing

- Screen printing

- Digital textile printing

Hand block Printing

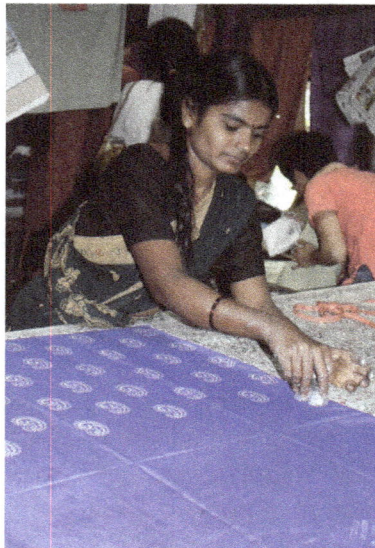

Woman doing block printing at Halasur village, Karnataka, India.

Wood handstamp for the textile printing of traditional paisley designs, Isfahan, Iran

This process is the earliest, simplest and slowest of all printing methods. A design is drawn on, or transferred to, prepared wooden blocks. A separate block is required for each distinct colour in the design. A blockcutter carves out the wood around the heavier masses first, leaving the finer and more delicate work until the last so as to avoid any risk of injuring it when the coarser parts are cut. When finished, the block has the appearance of a flat relief carving, with the design standing out. Fine details, difficult to cut in wood, are built up in strips of brass or copper, which is bent to shape and driven edgewise into the flat surface of the block. This method is known as coppering.

The printer applies colour to the block and presses it firmly and steadily on the cloth, striking it smartly on the back with a wooden mallet. The second impression is made in the same way, the printer taking care to see that it registers exactly with the first. Pins at each corner of the block join up exactly, so that the pattern can continue without a break. Each succeeding impression is made in precisely the same manner until the length of cloth is fully printed. The cloth is then wound over drying rollers. If the pattern contains several colours the cloth is first printed throughout with one color, dried, and then printed with the next.

Block printing by hand is a slow process. It is, however, capable of yielding highly artistic results, some of which are unobtainable by any other method. William Morris used this technique in some of his fabrics.

Perrotine Printing

The perrotine is a block-printing machine invented by Perrot of Rouen in 1834 and is now only of historical interest.

A Perrotine printing block

Roller, Cylinder, or Machine Printing

This process was patented by Bell in 1785, fifteen years after his use of an engraved plate to print textiles. Bell's patent was for a machine to print six colours at once, but, probably owing to its incomplete development, it was not immediately successful. One colour could be printed with satisfactorily; the difficulty was to keep the six rollers in register with each other. This defect was overcome by Adam Parkinson of Manchester in 1785. That year, Bells machine with Parkinson's improvement was successfully employed by Messrs Livesey, Hargreaves and Company of Bamber Bridge, Preston, for the printing of calico in from two to six colours at a single operation.

Roller printing was highly productive, 10,000 to 12,000 yards being commonly printed in one day of ten hours by a single-colour machine. It is capable of reproducing every style of design, ranging from the fine delicate lines of copperplate engraving to the small repeats and limited colours of the perrotine to the broadest effects of block printing with repeats from 1 in to 80 inches. It is precise, so each portion of an elaborate multicolour pattern can be fitted into its proper place without faulty joints at the points of repetition.

Stencil Printing

The art of stenciling on textile fabrics has been practised from time immemorial by the Japanese, and found increasing employment in Europe for certain classes of decorative work on woven goods during the late 19th century.

A pattern is cut from a sheet of stout paper or thin metal with a sharp-pointed knife, the uncut portions representing the part that will be left uncoloured. The sheet is laid on the fabric and colour is brushed through its interstices.

The peculiarity of stenciled patterns is that they have to be held together by ties. For instance, a complete circle cannot be cut without its centre dropping out, so its outline has to be interrupted at convenient points by ties or uncut portions. This limitation influences the design.

For single-colour work a stenciling machine was patented in 1894 by S. H. Sharp. It consists of an endless stencil plate of thin sheet steel that passes continuously over a revolving cast iron cylinder. The cloth to be ornamented passes between the two and the colour is forced onto it through the holes in the stencil by mechanical means.

Screen-printing

Screen printing is by far the most common technology today. Two types exist: rotary screen printing and flat (bed) screen printing. A blade (squeegee) squeezes the printing paste through openings in the screen onto the fabric.

Digital Textile Printing

Digital textile printing is often referred to as direct-to-garment printing, DTG printing, or digital garment printing. It is a process of printing on textiles and garments using specialized or modified inkjet technology. Inkjet printing on fabric is also possible with an inkjet printer by using fabric sheets with a removable paper backing. Today, major inkjet technology manufacturers can offer specialized products designed for direct printing on textiles, not only for sampling but also for bulk production. Since the early 1990s, inkjet technology and specially developed water-based ink (known as dye-sublimation or disperse direct ink) have made it possible to print directly onto polyester fabric. This is mainly related to visual communication in retail and brand promotion (flags, banners and other point of sales applications). Printing onto nylon and silk can be done by using an acid ink. Reactive ink is used for cellulose based fibers such as cotton and linen. Inkjet technology in digital textile printing allows for single pieces, mid-run production and even long-run alternatives to screen printed fabric.

Other Methods of Printing

Although most work is executed throughout by one or another of the seven distinct processes mentioned above, combinations are frequently employed. Sometimes a pattern is printed partly by machine and partly by block, and sometimes a cylindrical block is used along with engraved copper-rollers in an ordinary printing machine. The block in this latter case is in all respects, except for shape, identical with a flat wood or coppered block, but, instead of being dipped in colour, it receives its supply from an endless blanket, one part of which works in contact with colour-furnishing rollers and the other part with the cylindrical block. This block is known as a surface or peg roller. Many attempts have been made to print multicolour patterns with surface rollers alone, but hitherto with little success, owing to their irregularity in action and to the difficulty of preventing them from warping. These defects are not present in the printing of linoleum in which opaque oil colours are used, colours that neither sink into the body of the hard linoleum nor tend to warp the roller.

Lithographic printing has been applied to textile fabrics with qualified success. Its irregularity and the difficulty of registering repeats have restricted its use to the production of decorative panels, equal or smaller in size to the plate or stone.

Pad printing has been recently introduced to textile printing for the specific purpose of printing garment tags and care labels.

Calico Printing

Goods intended for calico printing are well-bleached; otherwise stains and other serious defects are certain to arise during subsequent operations.

The chemical preparations used for special styles will be mentioned in their proper places; but a general prepare, employed for most colours that are developed and fixed by steaming only, consists in passing the bleached calico through a weak solution of sulphated or turkey red oil containing 2.5 to 5 percent fatty acid. Some colours are printed on pure bleached cloth, but all patterns containing alizarine red, rose and salmon shades are considerably brightened by the presence of oil, and indeed very few, if any, colours are detrimentally affected by it.

The cloth is always brushed to free it from loose nap, flocks and dust that it picks up whilst stored. Frequently, too, it has to be sheared by being passed over rapidly revolving knives arranged spirally round an axle, which rapidly and effectually cuts off all filaments and knots, leaving the cloth perfectly smooth and clean. It is then stentered, wound onto a beam, and mounting on the printing machines.

Woollen Printing

The printing of wool.

Silk Printing

The colours and methods employed are the same as for wool, except that in the case of silk no preparation of the material is required before printing, and ordinary dry steaming is preferable to damp steaming.

Both acid and basic dyes play an important role in silk printing, which for the most part is confined to the production of articles for fashion goods, handkerchiefs, and scarves, all articles for which bright colours are in demand. Alizarine and other mordant colours are mainly used for any goods that have to resist repeated washings or prolonged exposure to light. In this case the silk frequently must be prepared in alizarine oil, after which it is treated in all respects like cotton, namely steamed, washed and soaped, the colours used being the same.

Silk is especially adapted to discharge and reserve effects. Most of the acid dyes can be discharged in the same way as when they are dyed on wool. Reserved effects are produced by printing mechanical resists, such as waxes and fats, on the cloth and then dyeing it in cold dye-liquor. The great affinity of the silk fibre for basic and acid dyestuffs enables it to extract colouring matter from cold solutions and permanently combine with it to form an insoluble lake. After dyeing, the reserve prints are washed, first in cold water to remove any colour not fixed onto the fibre, and then in hot water or benzene to dissolve out the resisting bodies.

After steaming, silk goods are normally only washed in hot water, but those printed entirely in mordant dyes will stand soaping, and indeed require it to brighten the colours and soften the material.

Some silk dyes do not require heat setting or steaming. They strike instantly, allowing the designer to dye color upon color. These dyes are intended mostly for silk scarf dyeing. They also dye bamboo, rayon, linen, and some other natural fabrics like hemp and wool to a lesser extent, but do not set on cotton.

Sanforization

A 1948 advertisement for sanforized cotton fabric

Sanforization is a process of treatment used for cotton fabrics mainly and most textiles made from natural or chemical fibres, patented by Sanford Lockwood Cluett (1874–1968) in 1930. It is a method of stretching, shrinking and fixing the woven cloth in both length and width before cutting and producing, to reduce the shrinkage which would otherwise occur after washing.

The cloth is continually fed into the sanforizing machine and therein moistened with either water or steam. A rotating cylinder presses a rubber sleeve against another, heated, rotating cylinder. Thereby the sleeve briefly gets compressed and laterally expanded, afterwards relaxing to its normal thickness. The cloth to be treated is transported between rubber sleeve and heated cylinder and is forced to follow this brief compression and lateral expansion, and relaxation. It is thus shrunk.

The greater the pressure applied to the rubber sleeve during the sanforization process, the less shrinking will occur once the shirt is in use. The process may be repeated.

The aim of the process is a cloth which does not shrink significantly during production by cutting, ironing, sewing or, especially, by wearing and washing the finished clothes. Cloth and articles made from it may be labelled to have a specific shrink-proof value (if pre-shrunk), e.g., of under 1%.

References

- Simpson, J. R.; Weiner, E. S. C. (1989). The Oxford English dictionary. Oxford: Clarendon Press. ISBN 0-19-861186-2.

- An Introduction to a History of Woodcut, Arthur M. Hind, p, Houghton Mifflin Co. 1935 (in USA), reprinted Dover Publications, 1963 ISBN 0-486-20952-0.

3

Formation of Textile

This chapter arms the reader with information regarding the processes that convert yarn to cloth or fabric. Weaving, knitting and crocheting have been studied in detail in the chapter. Such methods have been practiced for thousands of years. The chapter also talks about the machinery involved in these processes.

Weaving

Weaving is a method of textile production in which two distinct sets of yarns or threads are interlaced at right angles to form a fabric or cloth. Similar methods are knitting, felting, and braiding or plaiting. The longitudinal threads are called the warp and the lateral threads are the weft or filling. (*Weft* or *woof* is an old English word meaning "that which is woven".) The method in which these threads are inter woven affects the characteristics of the cloth.

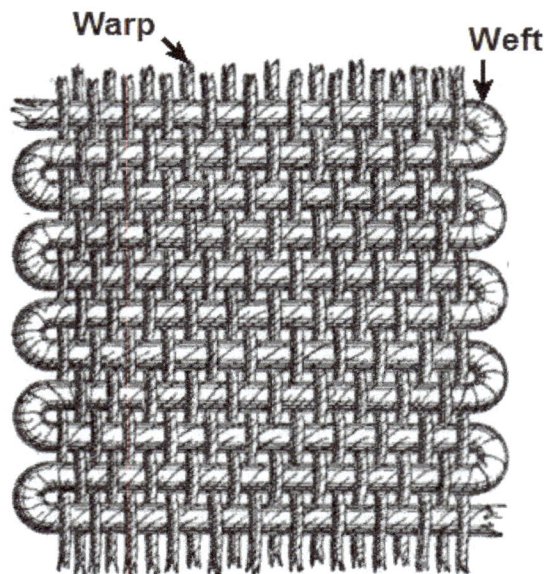

Warp and weft in plain weaving

Cloth is usually woven on a loom, a device that holds the warp threads in place while filling threads are woven through them. A fabric band which meets this definition of cloth (warp threads with a weft thread winding between) can also be made using other methods, including tablet weaving, back-strap, or other techniques without looms.

The way the warp and filling threads interlace with each other is called the weave. The majority of woven products are created with one of three basic weaves: plain weave, satin weave, or twill.

Woven cloth can be plain (in one colour or a simple pattern), or can be woven in decorative or artistic design.

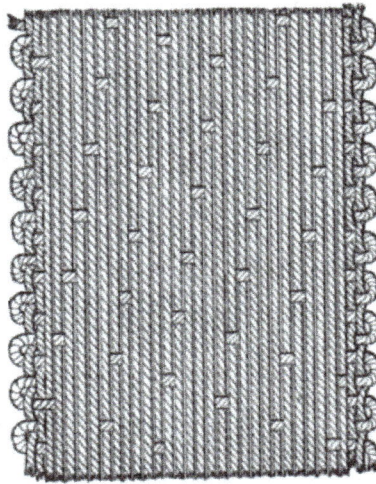

A satin weave, common for silk, each warp thread floats over 16 weft threads.

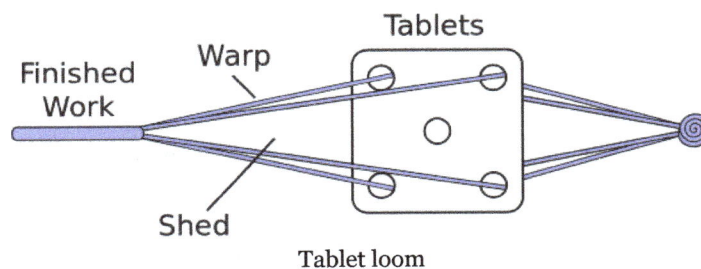

Tablet loom

Process and Terminology

In general, weaving involves using a loom to interlace two sets of threads at right angles to each other: the warp which runs longitudinally and the weft (older *woof*) that crosses it. One warp thread is called an end and one weft thread is called a pick. The warp threads are held taut and in parallel to each other, typically in a loom. There are many types of looms.

An Indian weaver preparing his warp on a pegged loom (another type of hand loom)
Patricia Hernandez Chavez and mother work on backstrap looms in Oaxaca

Free Standing Loom used commonly in Himachal Pradesh, India to weave shawls, caps etc.

Weaving can be summarized as a repetition of these three actions, also called the primary motion of the loom.

- Shedding: where the ends are separated by raising or lowering heald frames (heddles) to form a clear space where the pick can pass

- Picking: where the weft or pick is propelled across the loom by hand, an air-jet, a rapier or a shuttle.

- Beating-up or battening: where the weft is pushed up against the fell of the cloth by the reed.

The warp is divided into two overlapping groups, or lines (most often adjacent threads belonging to the opposite group) that run in two planes, one above another, so the shuttle can be passed between them in a straight motion. Then, the upper group is lowered by the loom mechanism, and the lower group is raised (shedding), allowing to pass the shuttle in the opposite direction, also in a straight motion. Repeating these actions form a fabric mesh but without beating-up, the final distance between the adjacent wefts would be irregular and far too large.

The secondary motion of the loom are the:

- Let off Motion: where the warp is let off the warp beam at a regulated speed to make the filling even and of the required design

- Take up Motion: Takes up the woven fabric in a regulated manner so that the density of filling is maintained

The tertiary motions of the loom are the stop motions: to stop the loom in the event of a thread break. The two main stop motions are the

- warp stop motion

- weft stop motion

The principal parts of a loom are the frame, the warp-beam or weavers beam, the cloth-roll (apron bar), the heddles, and their mounting, the reed. The warp-beam is a wooden or metal cylinder on the back of the loom on which the warp is delivered. The threads of the warp extend in parallel order from the warp-beam to the front of the loom where they are attached to the cloth-roll. Each thread or group of threads of the warp passes through an opening (eye) in a heddle. The warp threads are separated by the heddles into two or more groups, each controlled and automatically drawn up and down by the motion of the heddles. In the case of small patterns the movement of the heddles is controlled by "cams" which move up the heddles by means of a frame called a harness; in larger patterns the heddles are controlled by a dobby mechanism, where the healds are raised according to pegs inserted into a revolving drum. Where a complex design is required, the healds are raised by harness cords attached to a Jacquard machine. Every time the harness (the heddles) moves up or down, an opening (shed) is made between the threads of warp, through which the pick is inserted. Traditionally the weft thread is inserted by a shuttle.

On a conventional loom, the weft thread is carried on a pirn, in a shuttle that passes through the shed. A handloom weaver could propel the shuttle by throwing it from side to side with the aid of a picking stick. The "picking" on a power loom is done by rapidly hitting the shuttle from each side using an overpick or underpick mechanism controlled by cams 80-250 times a minute. When a pirn is depleted, it is ejected from the shuttle and replaced with the next pirn held in a battery attached to the loom. Multiple shuttle boxes allow more than one shuttle to be used. Each can carry a different colour which allows banding across the loom.

The rapier-type weaving machines do not have shuttles, they propel the weft by means of small grippers or rapiers that pick up the filling thread and carry it halfway across the loom where another rapier picks it up and pulls it the rest of the way. Some carry the filling yarns across the loom at rates in excess of 2,000 meters per minute. Manufacturers such as Picanol have reduced the mechanical adjustments to a minimum, and control all the functions through a computer with a graphical user interface. Other types use compressed air to insert the pick. They are all fast, versatile and quiet.

The warp is sized in a starch mixture for smoother running. The loom warped (loomed or dressed) by passing the sized warp threads through two or more heddles attached to harnesses. The power weavers loom is warped by separate workers. Most looms used for industrial purposes have a machine that ties new warps threads to the waste of previously used warps threads, while still on the loom, then an operator rolls the old and new threads back on the warp beam. The harnesses are controlled by cams, dobbies or a Jacquard head.

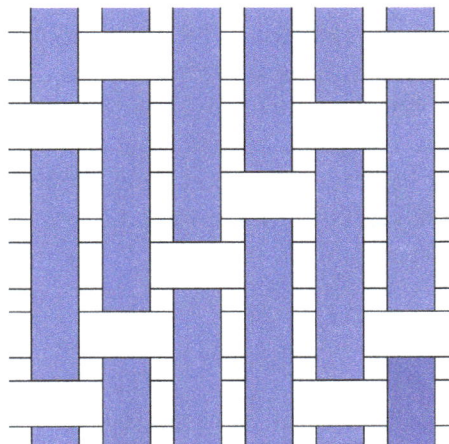

A 3/1 twill, as used in denim

The raising and lowering sequence of warp threads in various sequences gives rise to many possible weave structures: textile production went through profound changes brought about by the industrial revolution in the 19th century. At the beginning of the century in America, weaving was still done by hand, both commercially and at home. Most professional weavers were men who did their work for sale. Women wove items at home for family use. By the end of the 19th century weavers were simply mill workers who tended several water or steam powered looms at a time. The increased speed of production brought more textiles to the average farmhouse and renderd

- plain weave: plain, and hopsacks, poplin, taffeta, poult-de-soie, pibiones and grosgrain.

- twill weave: these are described by weft float followed by warp float, arranged to give diagonal pattern. 2/1 twill, 3/3 twill, 1/2 twill. These are softer fabrics than plain weaves.,

- satin weave: satins and sateens,

- complex computer-generated interlacings.

- pile fabrics : such as velvets and velveteens

Both warp and weft can be visible in the final product. By spacing the warp more closely, it can completely cover the weft that binds it, giving a *warp faced* textile such as repp weave. Conversely, if the warp is spread out, the weft can slide down and completely cover the warp, giving a *weft faced* textile, such as a tapestry or a Kilim rug. There are a variety of loom styles for hand weaving and tapestry.

History

Weaving in Ancient Egypt

There are some indications that weaving was already known in the Paleolithic era. An indistinct textile impression has been found at the Dolní Věstonice site.

The earliest known Neolithic textile production is supported by a 2013 find of a piece of cloth woven from hemp, in burial F. 7121 at the Çatalhöyük site suggested to be from around 7000 BC. Further finds come from the advanced civilisation preserved in the pile dwellings in Switzerland. Another extant fragment from the Neolithic was found in Fayum, at a site dated to about 5000 BCE. This fragment is woven at about 12 threads by 9 threads per cm in a plain weave. Flax was the predominant fibre in Egypt at this time (3600 BCE) and continued popularity in the Nile Valley, though wool became the primary fibre used in other cultures around 2000 BCE. Weaving was known in all the great civilisations, but no clear line of causality has been established. Early looms required two people to create the shed, and one person to pass through the filling. Early looms wove a fixed length of cloth, but later ones allowed warp to be wound out as the fell progressed. The weavers were often children or slaves. Weaving became simpler when the warp was sized.

The Americas

The oldest known textiles in the Americas are some fiberwork found in Guitarrero Cave, Peru dating back to 10,100 to 9,080 BCE.

In North America, pieces of 7,000- to 8,000-year-old fabric have been found with human burials at the Windover Archaeological Site in Florida. The burials were in a peat pond. The fabric had turned into peat, but was still identifiable. Many bodies at the site had been wrapped in fabric before burial. Eighty-seven pieces of fabric were found associated with 37 burials. Researchers have identified seven different weaves in the fabric. One kind of fabric had 26 strands per inch (10 strands per centimeter). There were also weaves using two-strand and three-strand wefts. A round bag made from twine was found, as well as matting. The yarn was probably made from palm leaves. Cabbage palm, saw palmetto and scrub palmetto are all common in the area, and would have been so 8,000 years ago.

Girls weaving a Persian rug in Hamadan, circa 1922. Note the design templates (called 'cartoons') at top of loom.

China and East Asia

The weaving of silk from silkworm cocoons has been known in China since about 3 500 BCE. Silk that was intricately woven and dyed, showing a well developed craft, has been found in a Chinese tomb dating back to 2 700 BCE.

Sericulture and silk weaving spread to Korea by 200 BCE, to Khotan by 50 CE, and to Japan by about 300 CE.

The pit-treadle loom may have originated in India though most authorities establish the invention in China. Pedals were added to operate heddles. By the Middle Ages such devices also appeared in Persia, Sudan, Egypt and possibly the Arabian Peninsula, where "the operator sat with his feet in a pit below a fairly low-slung loom." In 700 AD, horizontal looms and vertical looms could be found in many parts of Asia, Africa and Europe. In Africa, the rich dressed in cotton while the poorer wore wool. By the 12th century it had come to Europe either from the Byzantium or Moorish Spain where the mechanism was raised higher above the ground on a more substantial frame.

Medieval Europe

The predominant fibre was wool, followed by linen and nettlecloth for the lower classes. Cotton was introduced to Sicily and Spain in the 9th century. When Sicily was captured by the Normans, they took the technology to Northern Italy and then the rest of Europe. Silk fabric production was reintroduced towards the end of this period and the more sophisticated silk weaving techniques were applied to the other staples.

The weaver worked at home and marketed his cloth at fairs. Warp-weighted looms were commonplace in Europe before the introduction of horizontal looms in the 10th and 11th centuries. Weaving became an urban craft and to regulate their trade, craftsmen applied to establish a guild. These initially were merchant guilds, but developed into separate trade guilds for each skill. The cloth merchant who was a member of a city's weavers guild was allowed to sell cloth; he acted as a middleman between the tradesmen weavers and the purchaser. The trade guilds controlled quality and the training needed before an artisan could call himself a weaver.

Weaver, Nürnberg, c. 1425

By the 13th century, an organisational change took place, and a system of putting out was introduced. The cloth merchant purchased the wool and provided it to the weaver, who sold his produce back to the merchant. The merchant controlled the rates of pay and economically dominated the cloth industry. The merchants' prosperity is reflected in the wool towns of eastern England; Norwich, Bury St Edmunds and Lavenham being good examples. Wool was a political issue. The supply of thread has always limited the output of a weaver. About that time, the spindle method of spinning was replaced by the great wheel and soon after the treadle-driven spinning wheel. The loom remained the same but with the increased volume of thread it could be operated continuously.

The 14th century saw considerable flux in population. The 13th century had been a period of relative peace; Europe became overpopulated. Poor weather led to a series of poor harvests and starvation. There was great loss of life in the Hundred Years War. Then in 1346, Europe was struck with the Black Death and the population was reduced by up to a half. Arable land was labour-intensive and sufficient workers no longer could be found. Land prices dropped, and land was sold and put to sheep pasture. Traders from Florence and Bruges bought the wool, then sheep-owning landlords started to weave wool outside the jurisdiction of the city and trade guilds. The weavers started by working in their own homes then production was moved into purpose-built buildings. The working hours and the amount of work were regulated. The putting-out system had been replaced by a factory system.

Huguenot Weavers in the United Kingdom

Huguenot Weavers, Calvinists fleeing from religious persecution, migrated from mainland Europe to Britain around the time of 1685. They came from Flanders and major silk-weaving cities in France, such as Lyon and Tours. They settled first in Canterbury; then some 13,050 moved to Spitalfields in London. Their arrival had a major impact on the area economy, and Spitalfields consequently became known as "weaver town." Others moved further, to the silk weaving town of Macclesfield. Their arrival challenged the English weavers of cotton, woollen and worsted cloth, who subsequently learned the Huguenots' superior techniques. The influx of silk weavers greatly influenced the fashion tastes of the upper-class English, who began to incorporate more silk into their attire.

Weaving in the American Colonies (1500-1800)

Colonial America relied heavily on Great Britain for manufactured goods of all kinds. British policy was to encourage the production of raw materials in colonies and discourage manufacturing. The Wool Act 1699 restricted the export of colonial wool. As a result, many people wove cloth from locally produced fibres. The colonists also used wool, cotton and flax (linen) for weaving, though hemp could be made into serviceable canvas and heavy cloth. They could get one cotton crop each year; until the invention of the cotton gin it was a labour-intensive process to separate the seeds from the fibres.

A plain weave was preferred as the added skill and time required to make more complex weaves kept them from common use. Sometimes designs were woven into the fabric but most were added after weaving using wood block prints or embroidery.

Industrial Revolution

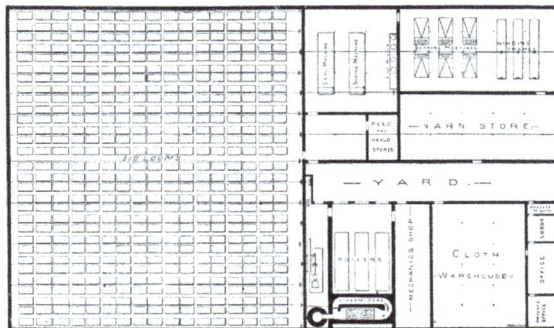

By 1892, most cotton weaving was done in similar weaving sheds, powered by steam.

Before the Industrial Revolution, weaving was a manual craft and wool was the principal staple. In the great wool districts a form of factory system had been introduced but in the uplands weavers worked from home on a putting-out system. The wooden looms of that time might be broad or narrow; broad looms were those too wide for the weaver to pass the shuttle through the shed, so that the weaver needed an expensive assistant (often an apprentice). This ceased to be necessary after John Kay invented the flying shuttle in 1733. The shuttle and the picking stick sped up the process of weaving. There was thus a shortage of thread or a surplus of weaving capacity. The opening of the Bridgewater Canal in June 1761 allowed cotton to be brought into Manchester, an area rich in fast flowing streams that could be used to power machinery. Spinning was the first to be mechanised (spinning jenny, spinning mule), and this led to limitless thread for the weaver.

Edmund Cartwright first proposed building a weaving machine that would function similar to recently developed cotton-spinning mills in 1784, drawing scorn from critics who said the weaving process was too nuanced to automate. He built a factory at Doncaster and obtained a series of patents between 1785 and 1792. In 1788, his brother Major John Cartwight built Revolution Mill at Retford (named for the centenary of the Glorious Revolution). In 1791, he licensed his loom to the Grimshaw brothers of Manchester, but their Knott Mill burnt down the following year (possibly a case of arson). Edmund Cartwight was granted a reward of £10,000 by Parliament for his efforts in 1809. However, success in power-weaving also required improvements by others, including H. Horrocks of Stockport. Only during the two decades after about 1805, did power-weaving take hold. At that time there were 250,000 hand weavers in the UK. Textile manufacture was one of the leading sectors in the British Industrial Revolution, but weaving was a comparatively late sector to be mechanised. The loom became semi-automatic in 1842 with Kenworthy and Bulloughs Lancashire Loom. The various innovations took weaving from a home-based artisan activity (labour-intensive and man-powered) to steam driven factories process. A large metal manufacturing industry grew to produce the looms, firms such as Howard & Bullough of Accrington, and Tweedales and Smalley and Platt Brothers. Most power weaving took place in weaving sheds, in small towns circling Greater Manchester away from the cotton spinning area. The earlier combination mills where spinning and weaving took place in adjacent buildings became rarer. Wool and worsted weaving took place in West Yorkshire and particular Bradford, here there were large factories such as Lister's or Drummond's, where all the processes took place. Both men and women with weaving skills emigrated, and took the knowledge to their new homes in New England, to places like Pawtucket and Lowell.

Jacquard Loom

Woven 'grey cloth' was then sent to the finishers where it was bleached, dyed and printed. Natural dyes were originally used, with synthetic dyes coming in the second half of the 19th century. The need for these chemicals was an important factor in the development of the chemical industry.

The invention in France of the Jacquard loom in about 1803, enabled complicated patterned cloths to be woven, by using punched cards to determine which threads of coloured yarn should appear on the upper side of the cloth. The jacquard allowed individual control of each warp thread, row by row without repeating, so very complex patterns were suddenly feasible. Samples exist showing calligraphy, and woven copies of engravings. Jacquards could be attached to handlooms or powerlooms.

The Role of the Weaver

A distinction can be made between the role and lifestyle and status of a handloom weaver, and that of the powerloom weaver and craft weaver. The perceived threat of the power loom led to disquiet and industrial unrest. Well known protests movements such as the Luddites and the Chartists had hand loom weavers amongst their leaders. In the early 19th century power weaving became viable. Richard Guest in 1823 made a comparison of the productivity of power and hand loom weavers:

A very good Hand Weaver, a man twenty-five or thirty years of age, will weave two pieces of nine-eighths shirting per week, each twenty-four yards long, and containing one hundred and five shoots of weft in an inch, the reed of the cloth being a forty-four, Bolton count, and the warp and weft forty hanks to the pound, A Steam Loom Weaver, fifteen years of age, will in the same time weave seven similar pieces.

He then speculates about the wider economics of using powerloom weavers:

...it may very safely be said, that the work done in a Steam Factory containing two hundred Looms, would, if done by hand Weavers, find employment and support for a population of more than two thousand persons.

Hand Loom Weavers

Hand loom weavers were mainly men- due to the strength needed to batten. They worked from home sometimes in a well lit attic room. The women of the house would spin the thread they needed, and attend to finishing. Later women took to weaving, they obtained their thread from the spinning mill, and working as outworkers on a piecework contract. Over time competition from the power looms drove down the piece rate and they existed in increasing poverty.

Power Loom Weavers

Power loom workers were usually girls and young women. They had the security of fixed hours, and except in times of hardship, such as in the cotton famine, regular income. They were paid a wage and a piece work bonus. Even when working in a combined mill, weavers stuck together and enjoyed a tight-knit community. The women usually minded the four machines and kept the looms oiled and clean. They were assisted by 'little tenters', children on a fixed wage who ran errands and did small tasks. They learnt the job of the weaver by watching. Often they would be half timers, carrying a green card which teacher and overlookers would sign to say they had turned up at the mill in the morning and in the afternoon at the school. At fourteen or so they come full-time into the mill, and started by sharing looms with an experienced worker where it was important to learn quickly as they would both be on piece work. Serious problems with the loom were left to the tackler to sort out. He would inevitably be a man, as were usually the overlookers. The mill had its health and safety issues, there was a reason why the women tied their hair back with scarves. Inhaling cotton dust caused lung problems, and the noise was causing total hearing loss. Weavers would mee-maw as normal conversation was impossible. Weavers used to 'kiss the shutttle' that is suck thread though the eye of the shuttle- this left a foul taste in the mouth due to the oil which was also carcinogenic.

Craft Weavers

Arts and Crafts was an international design philosophy that originated in England and flourished between 1860 and 1910 (especially the second half of that period), continuing its influence until the 1930s. Instigated by the artist and writer William Morris (1834–1896) during the 1860s and inspired by the writings of John Ruskin (1819–1900), it had its earliest and most complete development in the British Isles but spread to Europe and North America. It was largely a reaction against mechanisation and the philosophy advocated of traditional craftsmanship using simple forms and often medieval, romantic or folk styles of decoration. Hand weaving was highly regard and taken up as a decorative art.

Other Cultures

Andean Civilizations

Example of weaving characteristic of Andean civilizations.

Inca tunic

Whereas European cloth-making generally created ornamentation through "suprastructural" means—by adding embroidery, ribbons, brocade, dyeing, and other elements onto the finished woven textile—pre-Columbian Andean weavers created elaborate cloth by focusing on "structural" designs involving manipulation of the warp and weft of the fabric itself. Andeans used "tapestry techniques; double-, triple- and quadruple-cloth techniques; gauze weaves; warp-patterned weaves; discontinuous warp or scaffold weaves; and plain weaves" among many other techniques, in addition to the suprastructural techniques listed above.

American Southwest

Textile weaving, using cotton dyed with pigments, was a dominant craft among pre-contact tribes of the American southwest, including various Pueblo peoples, the Zuni, and the Ute tribes. The first Spaniards to visit the region wrote about seeing Navajo blankets. With the introduction of Navajo-Churro sheep, the resulting woolen products have become very well known. By the 18th century the Navajo had begun to import yarn with their favorite color, Bayeta red. Using an upright loom, the Navajos wove blankets worn as garments and then rugs after the 1880s for trade. Navajo traded for commercial wool, such as Germantown, imported from Pennsylvania. Under the influence of European-American settlers at trading posts, Navajos created new and distinct styles, including "Two Gray Hills" (predominantly black and white, with traditional patterns), "Teec Nos Pos" (colorful, with very extensive patterns), "Ganado" (founded by Don Lorenzo Hubbell), red dominated patterns with black and white, "Crystal" (founded by J. B. Moore), Oriental and Persian styles (almost always with natural dyes), "Wide Ruins," "Chinlee," banded geometric patterns, "Klagetoh," diamond type patterns, "Red Mesa" and bold diamond patterns. Many of these patterns exhibit a fourfold symmetry, which is thought to embody traditional ideas about harmony, or *hózhó*.

Weaving a traditional Navajo rug

Amazonia

In Native Amazonia, densely woven palm-bast mosquito netting, or tents, were utilized by the Panoans, Tupinambá, Western Tucano, Yameo, Záparoans, and perhaps by the indigenous peoples of the central Huallaga River basin (Steward 1963:520). Aguaje palm-bast (Mauritia flexuosa, Mauritia minor, or swamp palm) and the frond spears of the Chambira palm (Astrocaryum chambira, A.munbaca, A.tucuma, also known as Cumare or Tucum) have been used for centuries by the Urarina of the Peruvian Amazon to make cordage, net-bags hammocks, and to weave fabric. Among the Urarina, the production of woven palm-fiber goods is imbued with varying degrees of an aesthetic attitude, which draws its authentication from referencing the Urarina's primordial past. Urarina mythology attests to the centrality of weaving and its role in engendering Urarina society. The post-diluvial creation myth accords women's weaving knowledge a pivotal role in Urarina social reproduction. Even though palm-fiber cloth is regularly removed from circulation through mortuary rites, Urarina palm-fiber wealth is neither completely inalienable, nor fungible since it is a fundamental medium for the expression of labor and exchange. The circulation of palm-fiber wealth stabilizes a host of social relationships, ranging from marriage and fictive kinship (*compadrazco*, spiritual compeership) to perpetuating relationships with the deceased.

Crochet

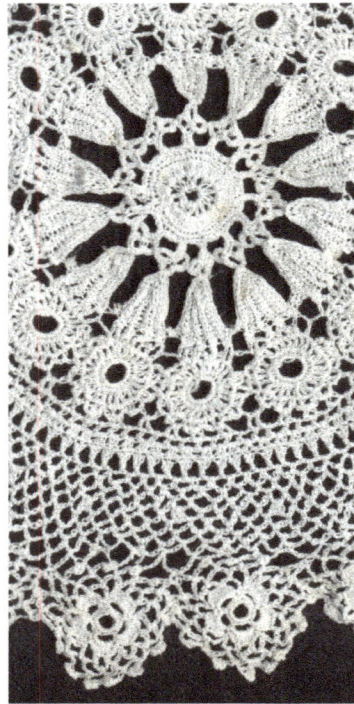

Detail of a crocheted doily, Sweden

Crochet is a process of creating fabric by interlocking loops of yarn, thread, or strands of other materials using a crochet hook. The name is derived from the French term "crochet", meaning *small hook*. These are made of materials such as metal, wood, or plastic and are manufactured commercially and produced in artisan workshops. The salient difference between crochet and knitting, beyond the implements used for their production, is that each stitch in crochet is completed before proceeding with the next one, while knitting keeps a large number of stitches open at a time. (Variant forms such as Tunisian crochet and broomstick lace keep multiple crochet stitches open at a time.)

Etymology

The word crochet is derived from the Old French *crochet*, a diminutive of *croche*, in turn from the Germanic *croc*, both meaning "hook". It was used in 17th-century French lace making, *crochetage* designating a stitch used to join separate pieces of lace, and *crochet* subsequently designating both a specific type of fabric and the hooked needle used to produce it. Although that fabric is not known to be crochet in the present sense, a genealogical relationship between the techniques sharing that name appears likely.

Origins

Knitted textiles survive from early periods but the first substantive evidence of crocheted fabric relates to its appearance in Europe during the 19th century. Earlier work identified as crochet was commonly made by nålebinding, a separate looped yarn technique.

A crocheted purse described in 1823 in *Penélopé*.

The first known published instructions for crochet appeared in the Dutch magazine, *Penélopé*, in 1823. This includes a color plate showing five different style purses of which three were intended to be crocheted with silk thread. The first is "simple open crochet" (*crochet simple ajour*), a mesh of chain-stitch arches. The second (illustrated here) starts in a semi-open form (*demi jour*), where chain-stitch arches alternate with equally long segments of slip-stitch crochet, and closes with a star made with "double-crochet stitches" (*dubbelde hekelsteek*— double-crochet in British terminology; single-crochet in US). The third purse is made entirely in double-crochet. The instructions prescribe the use of a tambour needle (as illustrated below) and introduce a number of decorative techniques.

The earliest dated English reference to garments made of cloth produced by looping yarn with a hook — *shepherd's knitting* — is in, *The Memoirs of a Highland Lady*, by Elizabeth Grant (1797–1830). The journal entry, itself, is dated 1812 but was not recorded in its subsequently published form until some time between 1845 and 1867, and the actual date of publication was first in 1898. Nonetheless, the 1833 volume of *Penélopé* describes and illustrates a shepherd's hook, and recommends its use for crochet with coarser yarn.

In 1842, one of the numerous books discussing crochet that began to appear in the 1840s states:

> "Crochet needles, sometimes called *Shepherds' hooks*, are made of steel, ivory, or box-wood. They have a hook at one end similar in shape to a fish-hook, by which the wool or silk is caught and drawn through the work. These instruments are to be procured of various sizes..."

Two years later, the same author, writes:

> "Crochet, — a species of knitting originally practised by the peasants in Scotland, with a small hooked needle called a shepherd's hook, — has, within the last seven years, aided by taste and fashion, obtained the preference over all other ornamental works of a similar nature. It derives its present name from the French; the instrument with which it is worked being by them, from its crooked shape, termed 'crochet.' This art has attained its highest degree of perfection in England, whence it has been transplanted to France and Germany, and both countries, although unjustifiably, have claimed the invention."

An instruction book from 1846 describes *Shepherd or Single Crochet* as what in current British usage is either called single crochet or slip-stitch crochet, with U.S. American terminology always using the latter (reserving single crochet for use as noted above). It similarly equates "Double" and "French crochet".

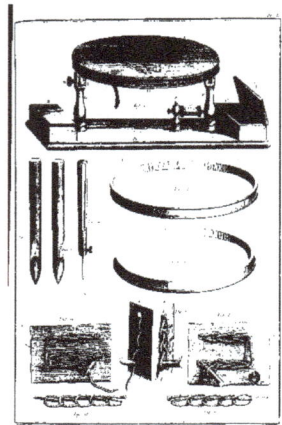

Tambour embroidery in the Diderot Encyclopedia

Notwithstanding the categorical assertion of a purely British origin, there is solid evidence of a connection between French tambour embroidery and crochet. The former method of production was illustrated in detail in 1763 in Diderot's Encyclopedia. The tip of the needle shown there is indistinguishable from that of a present-day inline crochet hook and the chain stitch separated from a cloth support is a fundamental element of the latter technique. The 1824 *Penélopé* instructions unequivocally state that the tambour tool was used for crochet and the first of the 1840s instruction books uses the terms *tambour* and *crochet* as synonyms. This equivalence is retained in the 4th edition of that work, 1847.

Shepherd's hook, 19th-century tapered hook, modern inline hook

The strong taper of the shepherd's hook eases the production of slip-stitch crochet but is less amenable to stitches that require multiple loops on the hook at the same time. Early yarn hooks were also continuously tapered but gradually enough to accommodate multiple loops. The design with a cylindrical shaft that is commonplace today was largely reserved for tambour-style steel needles. Both types gradually merged into the modern form that appeared toward the end of the 19th century, including both tapered and cylindrical segments, and the continuously tapered bone hook remained in industrial production until World War II.

The early instruction books make frequent reference to the alternate use of 'ivory, bone, or wooden hooks' and 'steel needles in a handle', as appropriate to the stitch being made. Taken with the synonymous labeling of shepherd's- and single crochet, and the similar equivalence of French- and double crochet, there is a strong suggestion that crochet is rooted both in tambour embroidery and shepherd's knitting, leading to thread and yarn crochet respectively; a distinction that is still made. The locus of the fusion of all these elements — the "invention" noted above — has yet to be determined, as does the origin of shepherd's knitting.

Shepherd's hooks are still being made for local slip-stitch crochet traditions. The form in the accompanying photograph is typical for contemporary production. A longer continuously tapering design intermediate between it and the 19th-century tapered hook was also in earlier production, commonly being made from the handles of forks and spoons.

Irish Crochet

Detail of a Portuguese crochet table-cloth, about 1970

Irish crochet lace, late 19th century. The design of this example is closely based on Flemish needle lace of the 17th century.

In the 19th century, as Ireland was facing the Great Irish Famine (1845-1849), crochet lace work was introduced as a form of famine relief (the production of crocheted lace being an alternative way of making money for impoverished Irish workers). Mademoiselle Riego de la Blanchardiere is generally credited with the invention of Irish Crochet, publishing the first book of patterns in 1846. Irish lace became popular in Europe and America, and was made in quantity until the first World War.

Modern Practice and Culture

Fashions in crochet changed with the end of the Victorian era in the 1890s. Crocheted laces in the new Edwardian era, peaking between 1910 and 1920, became even more elaborate in texture and complicated stitching.

Filet crochet by an internee at Manzanar War Relocation Center, 1943. Photograph by Ansel Adams

The strong Victorian colours disappeared, though, and new publications called for white or pale threads, except for fancy purses, which were often crocheted of brightly colored silk and elaborately beaded. After World War I, far fewer crochet patterns were published, and most of them were simplified versions of the early 20th-century patterns. After World War II, from the late 1940s until the early 1960s, there was a resurgence in interest in home crafts, particularly in the United States, with many new and imaginative crochet designs published for colorful doilies, potholders, and other home items, along with updates of earlier publications. These patterns called for thicker threads and yarns than in earlier patterns and included wonderful variegated colors. The craft remained primarily a homemaker's art until the late 1960s and early 1970s, when the new generation picked up on crochet and popularized granny squares, a motif worked in the round and incorporating bright colors.

Although crochet underwent a subsequent decline in popularity, the early 21st century has seen a revival of interest in handcrafts and DIY, as well as great strides in improvement of the quality and varieties of yarn. There are many more new pattern books with modern patterns being printed, and most yarn stores now offer crochet lessons in addition to the traditional knitting lessons. There are many books you can purchase from local book stores to teach yourself how to crochet whether it be as a beginner or intermediate. There are also many books for children and teenagers who are hoping to take up the hobby. Filet crochet, Tunisian crochet, tapestry crochet, broomstick lace, hairpin lace, cro-hooking, and Irish crochet are all variants of the basic crochet method.

Crochet has experienced a revival on the catwalk as well. Christopher Kane's Fall 2011 Ready-to-Wear collection makes intensive use of the granny square, one of the most basic of crochet motifs. In addition, crochet has been utilized many times by designers on the popular reality show *Project Runway*. Even websites such as Etsy and Ravelry have made it easier for individual hobbyists to sell and distribute their patterns or projects across the internet.

Bags and hacky sack tapestry crocheted in Guatemala.

TEJN sculpture dressed in crochet.

Laneya Wiles released a music video titled "Straight Hookin'" which makes a play on the word "hookers," which has a double meaning for both "one who crochets" and "a prostitute."

Materials

Basic materials required for crochet are a hook and some type of material that will be crocheted, most commonly yarn or thread. Additional tools are convenient for keeping stitches counted, measuring crocheted fabric, or making related accessories. Examples include cardboard cutouts, which can be used to make tassels, fringe, and many other items; a pom-pom circle, used to make pom-poms; a tape measure and a gauge measure, both used for measuring crocheted work and counting stitches; a row counter; and occasionally plastic rings, which are used for special projects. In recent years, yarn selections have moved beyond synthetic and plant and animal-based fibers to include bamboo, qiviut, hemp, and banana stalks, to name a few.

Crochet Hook

The crochet hook comes in many sizes and materials, such as bone, bamboo, aluminium, plastic, and steel. Because sizing is categorized by the diameter of the hook's shaft, a crafter aims to create stitches of a certain size in order to reach a particular gauge specified in a given pattern. If gauge is not reached with one hook, another is used until the stitches made are the needed size. Crafters may have a preference for one type of hook material over another due to aesthetic appeal, yarn glide, or hand disorders such as arthritis, where bamboo or wood hooks are favored over metal for the perceived warmth and flexibility during use. Hook grips and ergonomic hook handles are also available to assist crafters.

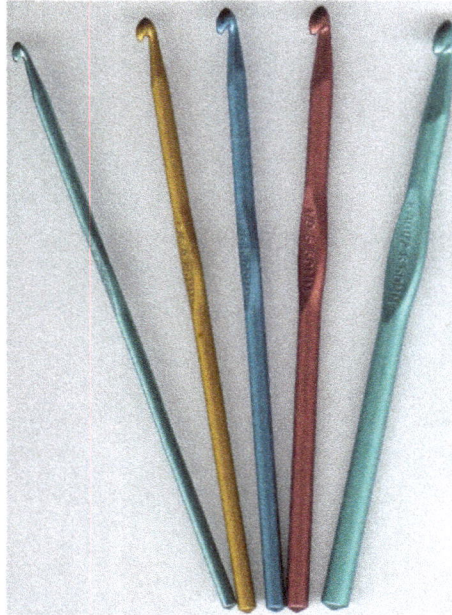

Aluminium crochet hooks

Steel crochet hooks range in size from 0.4 to 3.5 millimeters, or from 00 to 16 in American sizing. These hooks are used for fine crochet work such as doilies and lace.

Aluminium, bamboo, and plastic crochet hooks are available from 2.5 to 19 millimeters in size, or from B to S in American sizing.

Artisan-made hooks are often made of hand-turned woods, sometimes decorated with semi-precious stones or beads.

Crochet hooks used for Tunisian crochet are elongated and have a stopper at the end of the handle, while double-ended crochet hooks have a hook on both ends of the handle. There is also a double hooked apparatus called a Cro-hook that has become popular.

A hairpin loom is often used to create lacy and long stitches, known as hairpin lace. While this is not in itself a hook, it is a device used in conjunction with a crochet hook to produce stitches.

List of United States standard crochet hook and knitting needle sizes

Yarn

Yarn for crochet is usually sold as balls or skeins (hanks), although it may also be wound on spools or cones. Skeins and balls are generally sold with a *yarn band*, a label that describes the yarn's weight, length, dye lot, fiber content, washing instructions, suggested needle size, likely gauge, etc.

It is a common practice to save the yarn band for future reference, especially if additional skeins must be purchased. Crocheters generally ensure that the yarn for a project comes from a single dye lot. The dye lot specifies a group of skeins that were dyed together and thus have precisely the same color; skeins from different dye lots, even if very similar in color, are usually slightly different and may produce a visible stripe when added onto existing work. If insufficient yarn of a single dye lot is bought to complete a project, additional skeins of the same dye lot can sometimes be obtained from other yarn stores or online.

A hank of wool yarn (center) is uncoiled into its basic loop. A tie is visible at the left; after untying, the hank may be wound into a ball or balls suitable for crocheting. Crocheting from a normal hank directly is likely to tangle the yarn, producing snarls.

The thickness or weight of the yarn is a significant factor in determining the gauge, i.e., how many stitches and rows are required to cover a given area for a given stitch pattern. Thicker yarns generally require large-diameter crochet hooks, whereas thinner yarns may be crocheted with thick or thin hooks. Hence, thicker yarns generally require fewer stitches, and therefore less time, to work up a given project. Patterns and motifs are coarser with thicker yarns and produce bold visual effects, whereas thinner yarns are best for refined or delicate patternwork. Yarns are standardly grouped by thickness into six categories: superfine, fine, light, medium, bulky and superbulky. Quantitatively, thickness is measured by the number of wraps per inch (WPI). The related *weight per unit length* is usually measured in tex or denier.

Transformation of a hank of lavender silk yarn (top) into a ball in which the yarn emerges from the center (bottom). Using the latter is better for needlework, since the yarn is much less likely to tangle.

Before use, hanks are wound into balls in which the yarn emerges from the center, making crocheting easier by preventing the yarn from becoming easily tangled. The winding process may be performed by hand or done with a ballwinder and swift.

A yarn's usefulness is judged by several factors, such as its *loft* (its ability to trap air), its *resilience* (elasticity under tension), its washability and colorfastness, its *hand* (its feel, particularly softness vs. scratchiness), its durability against abrasion, its resistance to pilling, its *hairiness* (fuzziness), its tendency to twist or untwist, its overall weight and drape, its blocking and felting qualities, its comfort (breathability, moisture absorption, wicking properties) and its appearance, which includes its color, sheen, smoothness and ornamental features. Other factors include allergenicity, speed of drying, resistance to chemicals, moths, and mildew, melting point and flammability, retention of static electricity, and the propensity to accept dyes. Desirable properties may vary for different projects, so there is no one "best" yarn.

Although crochet may be done with ribbons, metal wire or more exotic filaments, most yarns are made by spinning fibers. In spinning, the fibers are twisted so that the yarn resists breaking under tension; the twisting may be done in either direction, resulting in a Z-twist or S-twist yarn. If the fibers are first aligned by combing them and the spinner uses a worsted type drafting method such as the short forward draw, the yarn is smoother and called a *worsted*; by contrast, if the fibers are carded but not combed and the spinner uses a woolen drafting method such as the long backward draw, the yarn is fuzzier and called *woolen-spun*. The fibers making up a yarn may be continuous *filament* fibers such as silk and many synthetics, or they may be *staples* (fibers of an average length, typically a few inches); naturally filament fibers are sometimes cut up into staples before spinning. The strength of the spun yarn against breaking is determined by the amount of twist, the length of the fibers and the thickness of the yarn. In general, yarns become stronger with more twist (also called *worst*), longer fibers and thicker yarns (more fibers); for example, thinner yarns require more twist than do thicker yarns to resist breaking under tension. The thickness of the yarn may vary along its length; a *slub* is a much thicker section in which a mass of fibers is incorporated into the yarn.

The spun fibers are generally divided into animal fibers, plant and synthetic fibers. These fiber types are chemically different, corresponding to proteins, carbohydrates and synthetic polymers, respectively. Animal fibers include silk, but generally are long hairs of animals such as sheep (wool), goat (angora, or cashmere goat), rabbit (angora), llama, alpaca, dog, cat, camel, yak, and muskox (qiviut). Plants used for fibers include cotton, flax (for linen), bamboo, ramie, hemp, jute, nettle, raffia, yucca, coconut husk, banana trees, soy and corn. Rayon and acetate fibers are also produced from cellulose mainly derived from trees. Common synthetic fibers include acrylics, polyesters such as dacron and ingeo, nylon and other polyamides, and olefins such as polypropylene. Of these types, wool is generally favored for crochet, chiefly owing to its superior elasticity, warmth and (sometimes) felting; however, wool is generally less convenient to clean and some people are allergic to it. It is also common to blend different fibers in the yarn, e.g., 85% alpaca and 15% silk. Even within a type of fiber, there can be great variety in the length and thickness of the fibers; for example, Merino wool and Egyptian cotton are favored because they produce exceptionally long, thin (fine) fibers for their type.

A single spun yarn may be crochet as is, or braided or plied with another. In plying, two or more yarns are spun together, almost always in the opposite sense from which they were spun individually; for example, two Z-twist yarns are usually plied with an S-twist. The opposing twist relieves some of the yarns' tendency to curl up and produces a thicker, *balanced* yarn. Plied yarns may themselves be plied together, producing *cabled yarns* or *multi-stranded yarns*. Sometimes, the yarns being plied are fed at different rates, so that one yarn loops around the other, as in bouclé. The single yarns may be dyed separately before plying, or afterwords to give the yarn a uniform look.

The dyeing of yarns is a complex art. Yarns need not be dyed; or they may be dyed one color, or a great variety of colors. Dyeing may be done industrially, by hand or even hand-painted onto the yarn. A great variety of synthetic dyes have been developed since the synthesis of indigo dye in the mid-19th century; however, natural dyes are also possible, although they are generally less brilliant. The color-scheme of a yarn is sometimes called its colorway. Variegated yarns can produce interesting visual effects, such as diagonal stripes.

Process

A close view of a crocheted scarf made with lace-weight mohair yarn.

Crocheted fabric is begun by placing a slip-knot loop on the hook (though other methods, such as a magic ring or simple folding over of the yarn may be used), pulling another loop through the first loop, and repeating this process to create a chain of a suitable length. The chain is either turned and worked in rows, or joined to the beginning of the row with a slip stitch and worked in rounds. Rounds can also be created by working many stitches into a single loop. Stitches are made by pulling one or more loops through each loop of the chain. At any one time at the end of a stitch, there is only one loop left on the hook. Tunisian crochet, however, draws all of the loops for an entire row onto a long hook before working them off one at a time. Like knitting, crochet can be worked either flat or in the round.

Types of Stitches

There are five main types of basic stitches. 1. Chain Stitch - the most basic of all stitches and used to begin most projects. 2. Slip Stitch - used to join chain stitch to form a ring. 3. Single Crochet Stitch - easiest stitch to master Single Crochet Stitch Tutorial 4. Half Double Crochet Stitch - the 'in-between' stitch Half-Double Crochet Tutorial 5. Double Crochet Stitch - many uses for this unlimited use stitch Double Crochet Stitch Tutorial

The more advanced stitches include the Shell Stitch, V Stitch, Spike Stitch, Afghan Stitch, Butterfly Stitch, Popcorn Stitch, and Crocodile Stitch.

International Crochet Terms and Notations

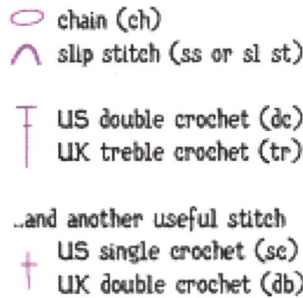

◯ chain (ch)
⋀ slip stitch (ss or sl st)

⊤ US double crochet (dc)
┃ UK treble crochet (tr)

...and another useful stitch
╷ US single crochet (sc)
┬ UK double crochet (db)

Some crochet symbols, abbreviations, and US/UK terms

In the English-speaking crochet world, basic stitches have different names that vary by country. The differences are usually referred to as UK/US or British/American. To help counter confusion when reading patterns, a diagramming system using a standard international notation has come into use (illustration, left).

Another terminological difference is known as *tension* (UK) and *gauge* (US). Individual crocheters work yarn with a loose or a tight hold and, if unmeasured, these differences can lead to significant size changes in finished garments that have the same number of stitches. In order to control for this inconsistency, printed crochet instructions include a standard for the number of stitches across a standard swatch of fabric. An individual crocheter begins work by producing a test swatch and compensating for any discrepancy by changing to a smaller or larger hook. North Americans call this *gauge*, referring to the end result of these adjustments; British crocheters speak of *tension*, which refers to the crafter's grip on the yarn while producing stitches.

Differences From and Similarities to Knitting

One of the more obvious differences is that crochet uses one hook while much knitting uses two needles. In most crochet, the artisan usually has only one live stitch on the hook (with the exception being Tunisian crochet), while a knitter keeps an entire row of stitches active simultaneously. Dropped stitches, which can unravel a fabric, rarely interfere with crochet work, due to a second structural difference between knitting and crochet. In knitting, each stitch is supported by the corresponding stitch in the row above and it supports the corresponding stitch in the row below, whereas crochet stitches are only supported by and support the stitches on either side of it. If a stitch in a finished crocheted item breaks, the stitches above and below remain intact, and because of the complex looping of each stitch, the stitches on either side are unlikely to come loose unless heavily stressed.

Round or cylindrical patterns are simple to produce with a regular crochet hook, but cylindrical knitting requires either a set of circular needles or three to five special double-ended needles. Many crocheted items are composed of individual motifs which are then joined together, either by sewing or crocheting, whereas knitting is usually composed of one fabric, such as entrelac.

Freeform crochet is a technique that can create interesting shapes in three dimensions because new stitches can be made independently of previous stitches almost anywhere in the crocheted piece. It is generally accomplished by building shapes or structural elements onto existing crocheted fabric at any place the crafter desires.

Knitting can be accomplished by machine, while many crochet stitches can only be crafted by hand. The height of knitted and crocheted stitches is also different: a single crochet stitch is twice the height of a knit stitch in the same yarn size and comparable diameter tools, and a double crochet stitch is about four times the height of a knit stitch.

While most crochet is made with a hook, there is also a method of crocheting with a knitting loom. This is called loomchet. Slip stitch crochet is very similar to knitting. Each stitch in slip stitch crochet is formed the same way as a knit or purl stitch which is then bound off. A person working in slip stitch crochet can follow a knitted pattern with knits, purls, and cables, and get a similar result.

It is a common perception that crochet produces a thicker fabric than knitting, tends to have less "give" than knitted fabric, and uses approximately a third more yarn for a comparable project than knitted items. Though this is true when comparing a single crochet swatch with a stockinette swatch, both made with the same size yarn and needle/hook, it is not necessarily true for crochet in general. Most crochet uses far less than 1/3 more yarn than knitting for comparable pieces, and a crocheter can get similar feel and drape to knitting by using a larger hook or thinner yarn. Tunisian crochet and slip stitch crochet can in some cases use less yarn than knitting for comparable pieces. According to sources claiming to have tested the 1/3 more yarn assertion, a single crochet stitch (sc) uses approximately the same amount of yarn as knit garter stitch, but more yarn than stockinette stitch. Any stitch using yarnovers uses less yarn than single crochet to produce the same amount of fabric. Cluster stitches, which are in fact multiple stitches worked together, will use the most length.

Standard crochet stitches like sc and dc also produce a thicker fabric, more like knit garter stitch. This is part of why they use more yarn. Slip stitch can produce a fabric much like stockinette that is thinner and therefore uses less yarn.

It is possible to use the same yarn or wool for both crochet and knitting, providing you have the correct size knitting needles or crochet hooks for the yarn you are using. There are some yarn that are only made for crochet, for example DMC make Cebelia No.10 which is a very thin yarn and works well with Amigurumi crochet.

- Differences between crochet and knitting

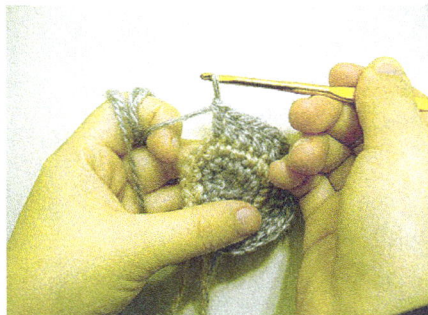

Most crochet uses one hook and works upon one stitch at a time. Crochet may be worked in circular rounds without any specialized tools, as shown here.

Knitting uses two or more straight needles that carry multiple stitches.

Unlike crochet, knitting requires specialized needles to create circular rounds.

Charity

It has been very common for people and groups to crochet clothing and other garments and then donate them to soldiers during war. People have also crocheted clothing and then donated it to hospitals, for sick patients and also for newborn babies. Sometimes groups will crochet for a specific charity purpose, such as crocheting for homeless shelters, nursing homes, etc. It is also becoming increasingly popular to crochet hats (commonly referred to as "chemo caps") and donate them to cancer treatment centers, for those undergoing chemotherapy. During the month of October pink hats and scarves are made and proceeds are donated to breast cancer funds.

Mathematics and Hyperbolic Crochet

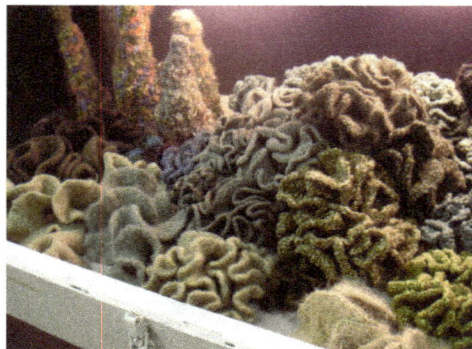

A collection of crocheted hyperbolic planes, in imitation of a coral reef.

Crochet has been used to illustrate shapes in hyperbolic space that are difficult to reproduce using other media or are difficult to understand when viewed two-dimensionally. A hyperbolic model of a coral reef has also been constructed for environmental purposes.

A paper model based on the pseudosphere was created by William Thurston, however, it was quite delicate. Crochet has been used by the mathematician Daina Taimina in order to create a version of the hyperbolic plane. Daina Taimina used the art of crochet to create a strong, durable model, which received an exhibition by the Institute For Figuring.

As hyperbolic and mathematics-based crochet has continued to become more popular, there have been several events highlighting work from various fiber artists. Two such shows include Sant Ocean Hall at the Smithsonian in Washington D.C. and Sticks, Hooks, and the Mobius: Knit and Crochet Go Cerebral at Lafayette College in Pennsylvania.

Architecture

In *Style in the technical arts*, Gottfried Semper looks at the textile with great promise and historical precedent. In Section 53, he writes of the "loop stitch, or Noeud Coulant: a knot that, if untied, causes the whole system to unravel." In the same section, Semper confesses his ignorance of the subject of crochet but believes strongly that it is a technique of great value as a textile technique and possibly something more.

There are a small number of architects currently interested in the subject of crochet as it relates to architecture. The following publications, explorations and thesis projects can be used as a resource to see how crochet is being used within the capacity of architecture.

- Emergent Explorations: Analog and Digital Scripting - Alexander Worden

- Research and Design: The Architecture of variation - Lars Spuybroek

- YurtAlert - Kate Pokorny

Yarn Bombing

In the past few years, a practice called yarn bombing, or the use of knitted or crocheted cloth to modify and beautify one's (usually outdoor) surroundings, emerged in the US and spread worldwide. Yarn bombers sometimes target existing pieces of graffiti for beautification. In 2010, an entity dubbed "the Midnight Knitter" hit West Cape May. Residents awoke to find knit cozies hugging tree branches and sign poles. In September 2015, Grace Brett was named "The World's Oldest Yarn Bomber". She is part of a group of yarn graffiti-artists called the Souter Stormers, who beautify their local town in Scotland. When she is not yarn bombing, she is utilizing her craft by making items for her children and grandchildren.

Entomology

Prolegs of lepidopteran larvae have a small circle of gripping hooks, called "crochets". The arrangement of the crochets can be helpful in identification to family level. Although the point has been debated, prolegs are not widely regarded as true legs, derived from the primitive uniramous limbs. Certainly in their morphology they are not jointed, and so lack the five segments (coxa, trochanter, femur, tibia, tarsus) of thoracic insect legs. Prolegs do have limited musculature, but much of their movement is hydraulically powered.

Felt

A selection of 4 different felt cloths

Kazakh felt yurt

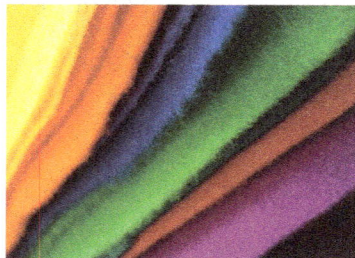

Samples of felt in different colours

Making a felt robe for Bakhtiari shepherds

Felt is a textile that is produced by matting, condensing and pressing fibres together. Felt can be made of natural fibres such as wool or synthetic fibres such as acrylic. There are many different types of felts for industrial, technical, designer and craft applications. While some types of felt are very soft, some are tough enough to form construction materials. Felt can vary in terms of fibre content, colour, size, thickness, density and more factors depending on the use of the felt.

Many cultures have legends as to the origins of felt making. Sumerian legend claims that the secret of feltmaking was discovered by Urnamman of Lagash. The story of Saint Clement and Saint Christopher relates that while fleeing from persecution, the men packed their sandals with wool to prevent blisters. At the end of their journey, the movement and sweat had turned the wool into felt socks.

Feltmaking is still practised by nomadic peoples (Altaic people: Mongols; Turkic people) in Central Asia, where rugs, tents and clothing are regularly made. Some of these are traditional items, such as the classic yurt (Gers), while others are designed for the tourist market, such as decorated slippers. In the Western world, felt is widely used as a medium for expression in textile art as well as design, where it has significance as an ecological textile.

Felting Methods

Wet Felting

Wet felting is one of several methods which can produce felt from wool and other animal fibres. Hot water is applied to layers of animal hairs. Repeated agitation and compression causes the fibres to hook together or weave together into a single piece of fabric. Wrapping the properly arranged fibre in a sturdy, textured material, such as a bamboo mat or burlap, will speed up the felting process. The felted material may be finished by fulling.

Only certain types of fibre can be wet felted successfully. Most types of fleece, such as those taken from the alpaca or the Merino sheep, can be put through the wet felting process. One may also use mohair (goat), angora (rabbit), or other rodent hair. These types of fibre are covered in tiny scales, similar to the scales found on a strand of human hair. Heat, motion, and moisture of the fleece causes the scales to open, while agitating them causes them to latch onto each other, creating felt. There is an alternate theory that the fibers wind around each other during felting. Plant fibres and synthetic fibres will not wet felt.

Needle Felting

Needle felting is a popular fibre arts craft that creates felt without the use of water. Special needles that are used in industrial felting machines are used by the artist as a sculpting tool. While erroneously referred to as "barbed" needles, they in fact have notches along the shaft of the needle that grab the top layer of fibres and tangle them with the inner layers of fibres as the needle enters the wool. Since these notches face down towards the tip of the needle, they do not pull the fibres out as the needle exits the wool, unless a reverse needle is used (with this desired effect). Once tangled and compressed using the needle, the felt can be strong and used for creating jewellery or sculpture. Finer details can be achieved with this method using a hand-held tool with either a single needle or a small group of needles (2-7), so it is a popular technique for producing 2D and 3D felted work.

Carroting

From the mid-17th to the mid-20th centuries, a process called "carroting" was used in the manufacture of good quality felt for making men's hats. Beaver, rabbit or hare skins were treated with a dilute solution of the mercury compound mercuric nitrate. The skins were dried in an oven where the thin fur at the sides turned orange, the colour of carrots. Pelts were stretched over a bar in a cutting machine, and the skin was sliced off in thin shreds, with the fleece coming away entirely. The fur was blown onto a cone-shaped colander and then treated with hot water to consolidate it. The cone then peeled off and passed through wet rollers to cause the fur to felt. These 'hoods' were then dyed and blocked to make hats. The toxic solutions from the carrot and the vapours it produced resulted in widespread cases of mercury poisoning among hatters.

Uses

Felt is used everywhere from the automotive industry and casinos, to musical instruments and home construction as felt paper. It is often used as a damper. In the automotive industry, for example, it damps the vibrations between interior panels and also stops dirt entering into some ball/cup joints. Felt is used on the underside of a car bra to protect the body. Gaskets are also often made of felt. Felt gaskets can be used for: seals, washers, spacers, stripping, discs, anti-vibration and anti-squeak pads, bumpers, bushings, noise reduction linings, vibration mounts, shock dampeners, heat barriers, wipers, oil and grease retainers, dust and fuel oil filters, sound deadening, padding, insulation, plugs, light seals, lamp bases, lubrication wicking, dust shields and filters.

Mosen felt textile,Tibet, 19th century. Similar textiles from western China were used as rugs. Shibori-dyed.

Many musical instruments use felt. On drum cymbal stands, it protects the cymbal from cracking and ensures a clean sound. It is used to wrap bass drum strikers and timpani mallets. Felt is used extensively in pianos; for example, piano hammers are made of wool felt around a wooden core. The density and springiness of the felt is a major part of what creates a piano's tone. As the felt becomes grooved and "packed" with use and age, the tone suffers. Felt is placed under the piano keys on accordions to control touch and key noise; it is also used on the *pallets* to silence notes not sounded by preventing air flow.

Felt is used for framing paintings. It is laid between the slip mount and picture as a protective measure to avoid damage from rubbing to the edge of the painting. This is commonly found as a preventive measure on paintings which have already been restored or professionally framed. It is widely used to protect paintings executed on various surfaces including canvas, wood panel and copper plate.

A felt-covered board can be used in storytelling to small children. Small felt cutouts or figures of animals, people, or other objects will adhere to a felt board, and in the process of telling the story, the storyteller also acts it out on the board with the animals or people. Puppets can also be made with felt. Felt pressed dolls were very popular in the nineteenth century and just after the first world war.

German artist Josef Beuys used felt in a number of works.

During the 18th and 19th centuries gentlemen's top hats made from beaver felt were quite popular. In the early part of the 20th century, cloth felt hats, such as fedoras, trilbies and homburgs, were worn by many men in the western world.

Bibliography

- E.J.W. Barber. *Prehistoric Textiles: The Development of Cloth in the Neolithic and Bronze Ages, with Special Reference to the Aegean*. Princeton: Princeton University Press, 1991.

- Lise Bender Jørgensen. *North European Textiles until AD 1000*. Aarchus: Aarchus University Press, 1992.

References

- Brown, Robin C. (1994). Florida's First People: 12,000 Years of Human History. Sarasota, Florida: Pineapple Press. p. 23. ISBN 1-56164-032-8.

- Milanich, Jerald T. (1994). Archaeology of Precolumbian Florida. Gainesville, Florida: University Press of Florida. pp. 74–75. ISBN 0-8130-1273-2.

- Broudy, Eric (1979). The Book of Looms: A History of the Handloom from Ancient Times to the Present. University Press of New England. pp. 111–112. ISBN 978-0874516494.

- Pacey, Arnold (1991), Technology in world civilization: a thousand-year history, MIT Press, pp. 40–1, ISBN 0-262-66072-5

- Jenkins, D.T., ed. (2003). The Cambridge History of Western Textiles, Volume 1. Cambridge University Press. p. 194. ISBN 978-0521341073.

- John A. Garraty; Mark C. Carnes (2000). "Chapter Three: America in the British Empire". A Short History of the American Nation (8th ed.). Longman. ISBN 0-321-07098-4.

- Campbell, Gordon (2006). The Grove Encyclopedia of Decorative Arts, Volume 1. Oxford University Press. ISBN 978-0-19-518948-3.

- Bartholomew Dean 2009 Urarina Society, Cosmology, and History in Peruvian Amazonia, Gainesville: University Press of Florida ISBN 978-0-8130-3378-5.

- Hodder, Ian (2013). "2013 Season Review" (PDF). Çatal Newsletter. pp. 1–2. Archived from the original (PDF) on 2015-04-13. Retrieved 7 February 2014.

- Stacey, Kevin. "Carbon dating identifies South America's oldest textiles." University of Chicago Press Journals. 13 April 2013.

4

Spinning (Textile) and Cotton Spinning Machinery

Spinning refers to the process of the conversion of cotton to yarn. This chapter gives an insight into both the olden and contemporary methods of spinning like mule spinning, ring spinning and break or open-end spinning. A section of this chapter is also dedicated to the array of machinery used in spinning like the spinning wheel, water frame, spinning mule, spinning jenny, throstle, ring frame and dref friction spinning. The major components of textile spinning are discussed in this chapter.

Spinning (Textiles)

Spinning is a major part of the textile industry. It is part of the textile manufacturing process where three types of fibre are converted into yarn, then fabrics, which undergo finishing processes such as bleaching to become textiles. The textiles are then fabricated into clothes or other products. There are three industrial processes available to spin yarn, and a handicraft community who use hand spinning techniques.

Spinning is the twisting together of drawn out strands of fibres to form yarn, though it is colloquially used to describe the process of drawing out, inserting the twist, and winding onto bobbins. In simple words, spinning is a process in which we convert fibers by passing through certain processes like Blow room, Carding, Drawing, Combing, Simplex, Ring Frame and finally winding into yarns. These yarns are then wound onto the cones.

Types of Fibre

Artificial fibres are made by extruding a polymer through a spinneret into a medium where it hardens. Wet spinning (rayon) uses a coagulating medium. In dry spinning (acetate and triacetate), the polymer is contained in a solvent that evaporates in the heated exit chamber. In melt spinning (nylons and polyesters) the extruded polymer is cooled in gas or air and sets. All these fibres will be of great length, often kilometers long.

Natural fibres are either from animals (sheep, goat, rabbit, silk-worm), mineral (asbestos), or from plants (cotton, flax, sisal). These vegetable fibres can come from the seed (cotton), the stem (known as bast fibres: flax, hemp, jute) or the leaf (sisal). Without exception, many processes are needed before a clean even staple is obtained – each with a specific name. With the exception of silk, each of these fibres is short, being only centimetres in length, and each has a rough surface that enables it to bond with similar staples.

Artificial fibres can be processed as long fibres or batched and cut so they can be processed like a natural fibre.

Spinning

Ring-spinning is the most common spinning method in the world. Other systems include air-jet and open-end spinning. Open-end spinning is done using break or open-end spinning. This is a technique where the staple fibre is blown by air into a rotor and attaches to the tail of formed yarn that is continually being drawn out of the chamber. Other methods of break spinning use needles and electrostatic forces.

The processes to make yarn short-staple yarn (typically spun from fibres from 0.75 to 2.0") are blending, opening, carding, pin-drafting, roving, spinning, and—if desired—plying and dyeing. In long staple spinning, the process may start with stretch-break of tow, a continuous "rope" of synthetic fibre. In open-end and air-jet spinning, the roving operation is eliminated. The spinning frame winds yarn a bobbin. Generally, after this step the yarn is wound to a cone for knitting or weaving.

> In mule spinning the roving is pulled off a bobbin and fed through rollers, which are feeding at several different speeds. This thins the roving at a consistent rate. If the roving was not a consistent size, then this step could cause a break in the yarn, or could jam the machine. The yarn is twisted through the spinning of the bobbin as the carriage moves out, and is rolled onto a cop as the carriage returns. Mule spinning produces a finer thread than the less skilled ring spinning.
>
> - The mule was an intermittent process, as the frame advanced and returned a distance of 5ft. It was the descendant of a 1779 Crompton device. It produces a softer, less twisted thread that was favoured for fines and for weft.
>
> - The ring was a descendant of the Arkwright water frame of 1769. It was a continuous process, the yarn was coarser, had a greater twist and was stronger so was suited to be warp. Ring spinning is slow due to the distance the thread must pass around the ring, and similar methods have improved on this; such as flyer and bobbin and cap spinning.
>
> Sewing thread, was made of several threads twisted together, or doubled.

The pre-industrial techniques of hand spinning with spindle or spinning wheel continue to be practiced as a handicraft or hobby, and enable wool or unusual vegetable and animal staples to be creatively used.

• Checking

> This is the process where each of the bobbins is rewound to give a tighter bobbin.

• Folding and Twisting

> Plying is done by pulling yarn from two or more bobbins and twisting it together, in the opposite direction from that in which it was spun. Depending on the weight desired, the yarn may or may not be plied, and the number of strands twisted together varies.

History and Economics

Hand-spinning was a cottage industry in medieval Europe, where the wool spinners (often women and children) would provide enough yarn to service the needs of the men who operated the loom.

This would occur in districts favourable to sheep husbandry. The introduction of the flying shuttle upset this balance. The subsequent invention of the spinning jenny water frame redressed the balance but required water power to operate the machinery, and the industry relocated to West Yorkshire where this was available. The nascent cotton industry was located on wetter side of the same hills. The British government was very protective of this technology, restricting its export. By the aftermath of World War I the colonies where the cotton was grown started to purchase and manufacture significant quantities of cotton spinning machinery. The next breakthrough was with the move over to break or open-end spinning, and then the adoption of artificial fibres. By then most production had moved to India and China.

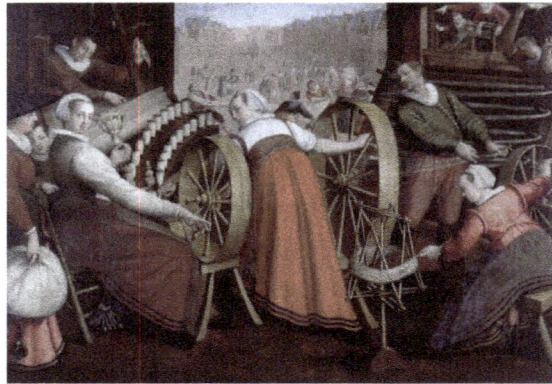

1595 painting illustrating Leiden textile workers

During the industrial revolution, spinners, doffers, and sweepers were employed in spinning mills for the 18th to 20th centuries. Many employees of the mill were children, who were preferred due to their small size and agility.

Cotton-spinning Machinery

Cotton-spinning machinery refers to machines which process (or spin) prepared cotton roving into workable yarn or thread. Such machinery can be dated back centuries. During the 18th and 19th centuries, as part of the Industrial Revolution cotton-spinning machinery was developed to bring mass production to the cotton industry. Cotton spinning machinery was installed in large factories, commonly known as cotton mills.

History

Until the 1740s all spinning was done by hand using a spinning wheel. The state of the art spinning wheel in England was known as the Jersey wheel however an alternative wheel, the Saxony wheel was a double band treadle spinning wheel where the spindle rotated faster than the traveller in a ratio of 8:6, drawing on both was done by the spinners fingers.

In 1738 Lewis Paul and John Wyatt of Birmingham patented the Roller Spinning machine and the flyer-and-bobbin system, for drawing cotton to a more even thickness, using two sets of rollers that travelled at different speeds. This principle was the basis of Richard Arkwright's later water frame design. By 1742 Paul and Wyatt had opened a mill in Birmingham which used their new rolling machine powered by a donkey, this was not profitable and soon closed. A factory was opened in Northampton in 1743, with fifty spindles turning on five of Paul and Wyatt's machines, proving more successful than their first mill; this operated until 1764.

Lewis Paul invented the hand-driven carding machine in 1748. A coat of wire slips were placed around a card, which was then wrapped around a cylinder. Lewis' invention was later developed and improved by Richard Arkwright and Samuel Crompton, although the design came under suspicion after a fire at Daniel Bourn's factory in Leominster which used Paul and Wyatt's spindles. Bourn produced a similar patent in the same year.

Old advertising display of items used in cotton textile manufacture during the industrial revolution

Rev John Dyer of Northampton recognised the importance of the Paul and Wyatt cotton spinning machine in a poem in 1757:

A circular machine, of new designIn conic shape: it draws and spins a threadWithout the tedious toil of needless hands.A wheel invisible, beneath the floor,To ev'ry member of th' harmonius frame,Gives necessary motion. One intentO'erlooks the work; the carded wool, he says,So smoothly lapped around those cylinders,Which gently turning, yield it to yon cirueOf upright spindles, which with rapid whirlSpin out in long extenet an even twine.

Spinning Jenny

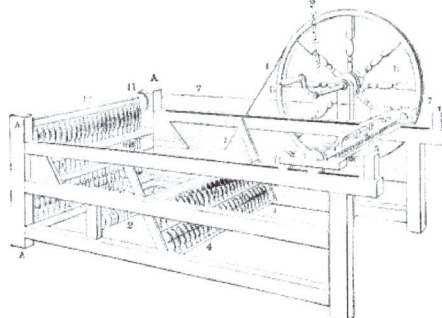

The spinning jenny that was used in textile mills

The spinning jenny is a multi-spool spinning wheel. It was invented circa 1764, its invention attributed to James Hargreaves in Stanhill, near Blackburn, Lancashire.

Water Frame

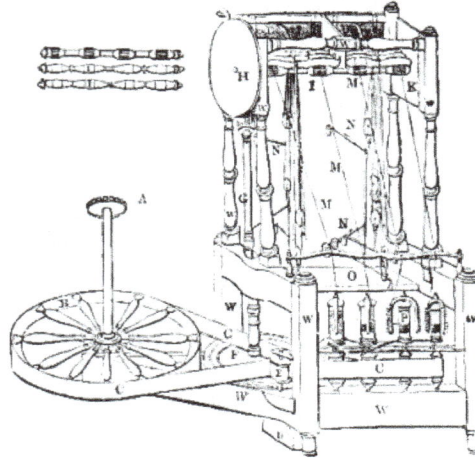

Arkwright's spinning frame

The Water frame was developed and patented by Arkwright in the 1770s. The roving was attenuated (stretched) by drafting rollers and twisted by winding it onto a spindle. It was heavy large scale machine that needed to be driven by power, which in the late 18th century meant by a water wheel. Cotton mills were designed for the purpose by Arkwright, Jedediah Strutt and others along the River Derwent in Derbyshire. Water frames could only spin weft.

Spinning Mule

A fully restored & working mule at the Quarry Bank Mill, UK.

The spinning mule or mule jenny was created in 1779 by Samuel Crompton. It was a combination of Arkwright's water frame and Hargreaves' spinning jenny. It was so named because it was a hybrid of these two machines. The mule consisted of a fixed frame containing a creel of bobbins holding the roving, connected through the headstock to a parallel carriage containing the spindles. It used an intermittent process: On the outward traverse, the rovings were paid out, and twisted, and the return traverse, the roving was clamped and the spindles reversed taking up the newly spun thread. The rival machine, the throstle frame or ring frame was a continuous process, where the roving was drawn twisted and wrapped in one action. The spinning mule became self-acting (automatic) in 1830s. The mule was the most common spinning machine from 1790 until about 1900, but was still used for fine yarns until the 1960s. A cotton mill in 1890 would contain over 60 mules, each with 1320 spindles.

Between the years 1824 and 1830 Richard Roberts invented a mechanism that rendered all parts of the mule self-acting, regulating the rotation of the spindles during the inward run of the carriage.

The Platt Brothers, based in Oldham, Greater Manchester were amongst the most prominent machine makers in this field of work.

At first this machine was only used to spin coarse and low-to-medium counts, but it is now employed to spin all counts of yarn.

Throstle

The Throstle frame was a descendant of the water frame. It used the same principles, was better engineered and driven by steam. In 1828 the Danforth throstle frame was invented in the United States. The heavy flyer caused the spindle to vibrate, and the yarn snarled every time the frame was stopped. Not a success. It was named throstle, as the noise it made when running was compared to the song of the throstle (thrush).

Ring Frame

The Ring frame is credited to John Thorp in Rhode Island in 1828/9 and developed by Mr. Jencks of Pawtucket, Rhode Island, who (Marsden 1885) names as the inventor.

The bobbins or tubes may be filled from "cops", "ring spools" or "hanks", but a stop motion is required for each thread, which will come into operation immediately a fracture occurs.

Break or Open-end Spinning

Dref Friction Spinning

Further Processes

For many purposes, the threads as spun by the ring frame or the mule are ready for the manufacturer; but where extra strength or smoothness is required, as in threads for sewing, crocheting, hosiery, lace and carpets; also where multicoloured effects are needed, as in Grandrelle, or some special form of irregularity, as in corkscrewed, and knopped yarns, two or more single threads are compounded and twisted together. This operation is known as "doubling".

Spinning Wheel

C spinning wheel is a device for spinning thread or yarn from natural or [[synthetic K Kfibre|synthetic]] fibres. Spinning wheels were first used in India, between 500 and 1000 C.E. Spinning machinery, such as the spinning jenny and spinning frame, displaced the spinning wheel during the Industrial Revolution.

Irish spinning wheel - around 1900 Library of Congress collection

HINDOO SPINNING-WHEEL.

Hindoo Spinning-Wheel (1852)

History

Detail of *The Spinning Wheel*, by Chinese artist Wang Juzheng, Northern Song Dynasty (960–1279)

The spinning wheel was invented in India, between 500 and 1000 C.E. The earliest clear illustrations of the spinning wheel come from Baghdad (drawn in 1234), China (*c.* 1270) and Europe (*c.* 1280), and there is evidence that spinning wheels had already come into use in both China [a] and the Islamic world during the eleventh century. In France the spindle and distaff were not displaced until the mid 18th century.

Scene from Al-Maqamat, painted by al-Wasiti 1237

The spinning wheel replaced the earlier method of hand spinning with a spindle. The first stage in mechanizing the process was mounting the spindle horizontally so it could be rotated by a cord encircling a large, hand-driven wheel. The great wheel is an example of this type, where the fibre is held in the left hand and the wheel slowly turned with the right. Holding the fibre at a slight angle to the spindle produced the necessary twist. The spun yarn was then wound onto the spindle by moving it so as to form a right angle with the spindle. This type of wheel, while known in Europe by the 14th century, was not in general use until later. The construction of the Great Wheel made it very good at creating long drawn soft fuzzy wools, but very difficult to create the strong smooth yarns needed to create warp for weaving. Spinning wheels ultimately did not develop the capability to spin a variety of yarns until the beginning of the 19th century and the mechanization of spinning.

Woman spinning with a wheel, from the Elizabethan era, early 17th century

In general, the spinning technology was known for a long time before being adopted by the majority of people, thus making it hard to fix dates of the improvements. In 1533, a citizen of Brunswick is said to have added a treadle, by which the spinner could rotate her spindle with one foot and have both hands free to spin. Leonardo da Vinci drew a picture of the flyer, which twists the yarn before winding it onto the spindle. During the 16th century a treadle wheel with flyer was in common use, and gained such names as the Saxony wheel and the flax wheel. It sped up production, as one needn't stop spinning to wind up the yarn.

Elder citizen of Suwałki region, Poland - Mrs. Skindzierz-Jakubowska (82) at the spinning wheel, 1976

On the eve of the Industrial revolution it took at least five spinners to supply one weaver. Lewis Paul and John Wyatt first worked on the problem in 1738, patenting the Roller Spinning machine and the flyer-and-bobbin system, for drawing wool to a more even thickness. Using two sets of rollers that travelled at different speeds, yarn could be twisted and spun quickly and efficiently. However, they did not have much financial success. In 1771, Richard Arkwright used waterwheels to power looms for the production of cotton cloth, his invention becoming known as the water frame.

More modern spinning machines use a mechanical means to rotate the spindle, as well as an automatic method to draw out fibres, and devices to work many spindles together at speeds previously unattainable. Newer technologies that offer even faster yarn production include friction spinning, an open-end system, and air jets.

Types

A depiction of spinning by Diego Rodríguez de Silva y Velázquez, 1644-1648

Numerous types of spinning wheels exist, including the *great wheel* also known as *walking wheel* or *wool wheel* for rapid long draw spinning of woolen-spun yarns; the *flax wheel*, which is a double-drive wheel used with a distaff for spinning linen; *saxony* and *upright* wheels, all-purpose treadle driven wheels used to spin both woolen and worsted-spun yarns; and the *charkha*, native to Asia. Until the acceptance of rotor spinning wheel, all yarns were produced by aligning fibres through drawing techniques and then twisting the fiber together. With rotor spinning, the fibers in the roving are separated, thus opened, and then wrapped and twisted as the yarn is drawn out of the rotor cup.

Charkha

The tabletop or floor charkha is one of the oldest known forms of the spinning wheel. The charkha works similarly to the great wheel, with a drive wheel being turned by hand, while the yarn is spun off the tip of the spindle. The floor charkha and the great wheel closely resemble each other. With both, the spinning must stop in order to wind the yarn onto the spindle.

The word *charkha* which has links with Persian چرخ: charKh, wheel, is related to the word "circle" (rather like "church" is to "kirk"). The charkha was both a tool and a symbol of the Indian independence movement. The charkha, a small, portable, hand-cranked wheel, is ideal for spinning cotton and other fine, short-staple fibres, though it can be used to spin other fibers as well. The size varies, from that of a hardbound novel to the size of a briefcase, to a floor charkha. Mahatma Gandhi brought the charkha into wider use with his teachings. He hoped the charkha would assist the people of India achieve self-sufficiency and independence, and therefore used the charkha as a symbol of the Indian independence movement and included it on earlier versions of the Flag of India.

Modified and portable compact Charkha

Great wheel

2Spinning wool on a great wheel at a demonstration in the Conner Prairie living history museum loom house

The great wheel was one of the earlier types of spinning wheel. The fiber is held in the left hand and the wheel slowly turned with the right. This wheel is thus good for using the long-draw spinning technique, which requires only one active hand most of the time, thus freeing a hand to turn the wheel. The great wheel is usually used to spin short-staple fibers (this includes both cotton and wool), and can only be used with fibre preparations that are suited to long-draw spinning.

The great wheel is usually over 5 feet (1.5 m) in height. The large drive wheel turns the much smaller spindle assembly, with the spindle revolving many times for each turn of the drive wheel. The yarn is spun at an angle off the tip of the spindle, and is then stored on the spindle. To begin spinning on a great wheel, first a leader (a length of waste yarn) is tied onto the base of the spindle and spiraled up to the tip. Then the spinner overlaps a handful of fiber with the leader, holding both gently together with the left hand, and begins to slowly turn the drive wheel clockwise with the right hand, while simultaneously walking backward and drawing the fibre in the left hand away from the spindle at an angle. The left hand must control the tension on the wool to produce an even result. Once a sufficient amount of yarn has been made, the spinner turns the wheel backward a short distance to unwind the spiral on the spindle, then turns it clockwise again, and winds the newly made yarn onto the spindle, finishing the wind-on by spiralling back out to the tip again to make another draw.

Treadle Wheel

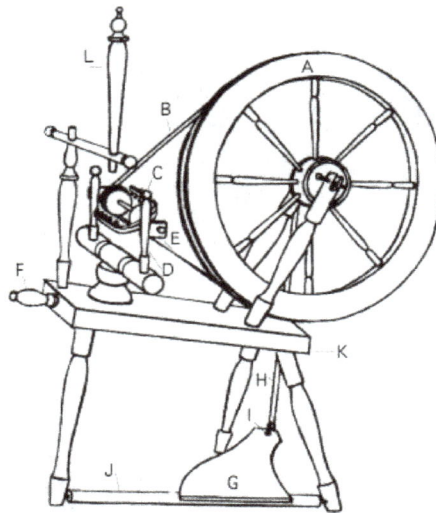

Parts of a treadle wheel: A - Wheel, B - Drive band, C - Flyer assembly, D - Maiden, E - Bearings, F - Tension Screw, G - Treadle, H - Footman, I - Treadle connection, J - Treadle bar, K - Table, L - Distaff

This type of wheel is powered by the spinner's foot rather than their hand or a motor. The spinner sits and pumps a foot treadle that turns the drive wheel via a crankshaft and a connecting rod. This leaves both hands free for drafting the fibres, which is necessary in the short draw spinning technique, which is often used on this type of wheel. The old-fashioned pointed driven spindle is not a common feature of the treadle wheel. Instead, most modern wheels employ a flyer-and-bobbin system which twists the yarn and winds it onto a spool simultaneously. These wheels can be single- or double-treadle; which is a matter of preference and does not affect the operation of the wheel.

Double Drive

A double drive wheel

The double drive wheel is named after its drive band, which goes around the spinning wheel twice. The drive band turns the flyer, which is the horse-shoe shaped piece of wood surrounding the bobbin, as well as the bobbin. Due to a difference in the size of the whorls (the round pieces or pulleys around which the drive band runs) the bobbin whorl, which has a smaller radius than the flyer whorl, turns slightly faster. Thus both the flyer and bobbin rotate to twist the yarn, and the difference in speed continually winds the yarn onto the bobbin. Generally the speed difference or "ratio" is adjusted by the size of the whorls and the tension of the drive band.

The drive band on the double drive wheel is generally made from a non-stretch yarn or twine; candlewick is also used.

Single Drive

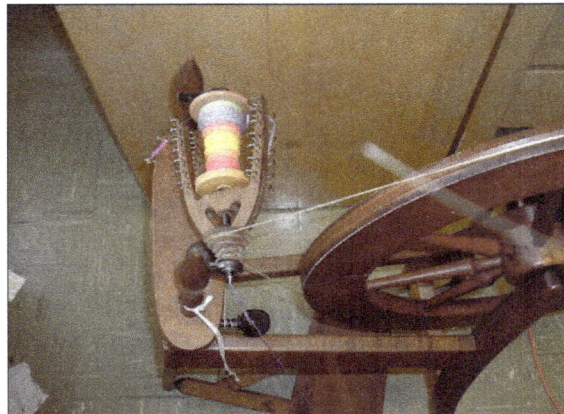

A single-drive wheel with the drive band around flyer and brake on the bobbin

A single drive wheel has one drive band that goes around both the flywheel and the flyer, and a short tension band which goes only over the bobbin. The tension band adds an adjustable amount of drag to the bobbin and thereby increases the yarn take up force.

If the tension band were extremely tight and the bobbin could not rotate at all, yarn would be taken up onto the bobbin by the rotation of the flyer constantly at a rate of one wrap per revolution of the flyer. In practice, the tension is set such that the bobbin can slip, but with some drag, generating the differential rate of rotation between the flyer and the bobbin. This drag is the force which winds new yarn onto the bobbin.

While the spinner is making new yarn, the bobbin and the flyer turn in unison, driven by the single drive band. When the spinner feeds the yarn onto the bobbin, the drag on the flyer slows it and thus the yarn winds on. The tighter the tension band is, the more pull on the yarn, because the more friction the bobbin has to overcome to turn in sync with the flyer.

Castle Style

When the spindle and flyer are located above the wheel, rather than off to one side, the wheel is said to be a castle wheel. This type of wheel is often more compact, thus easier to store. Some castle wheels are even made to fold up small enough that they fit in carry-on luggage at the airport.

Electric Spinning Wheel

Electric spinning wheels or e-spinners are powered by an electric motor rather than via a treadle. Some require mains power while others may be powered by a low-voltage source, such as a rechargeable battery. Most e-spinners are small and portable.

One of the attractions of an e-spinner is that it is not necessary to coordinate treadling with handling the fiber (drafting), so it is generally easier to learn to spin on an e-spinner than a traditional treadle-style spinning wheel. E-spinners are also suitable for spinners who have trouble treadling for various reasons.

E-spinners represent an evolution of the tools used in the craft of handspinning, similar to what has occurred in sewing, quilting, woodworking, and other crafts.

Importance

The spinning wheel increased the productivity of thread making by a factor of greater than 10. Medieval historian Lynn White credited the spinning wheel with increasing the supply of rags, which led to cheap paper, which was a factor in the development of printing.

Culture

The ubiquity of the spinning wheel has led to its inclusion in the art, literature and other expressions of numerous cultures around the world, and in the case of South Asia it has become a powerful political symbol.

Political Symbolism

Starting in 1931, the traditional spinning wheel became the primary symbol on the flag of the Provisional Government of Free India.

Mahatma Gandhi's manner of dress and commitment to hand spinning were essential elements of his philosophy and politics. He chose the traditional loincloth as a rejection of Western culture and a symbolic identification with the poor of India. His personal choice became a powerful political gesture as he urged his more privileged followers to copy his example and discard—or even burn—their European-style clothing and return with pride to their ancient, precolonial culture. Gandhi claimed that spinning thread in the traditional manner also had material advantages, as it would create the basis for economic independence and the possibility of survival for India's impoverished rural areas. This commitment to traditional cloth making was also part of a larger swadeshi movement, which aimed for the boycott of all British goods. As Gandhi explained to Charlie Chaplin in 1931, the return to spinning did not mean a rejection of all modern technology but of the exploitative and controlling economic and political system in which textile manufacture had become entangled. Gandhi said, "Machinery in the past has made us dependent on England, and the only way we can rid ourselves of the dependence is to boycott all goods made by machinery. This is why we have made it the patriotic duty of every Indian to spin his own cotton and weave his own cloth."

Mahatma Gandhi spinning yarn on a charkha.

Literature and Folk Tales

St. Elisabeth of Hungary spinning for the poor, a depiction of the castle style spinning wheel in art. Note also the distaff used to hold the fiber.

The Golden Spinning Wheel (Zlatý kolovrat) is a Czech poem by Karel Jaromír Erben that was included in his classic collection of folk ballads, Kytice.

Rumpelstiltskin, one of the tales collected by the Brothers Grimm, revolves around a woman who is imprisoned under threat of execution unless she can spin straw into gold. Rumpelstiltskin helps her with this task, ultimately at the cost of her first-born child; however, she makes a new bargain with him and is able to keep her child after successfully guessing his name.

Another folk tale that incorporates spinning wheels is the classic fairy tale *Sleeping Beauty*, in which the main character pricks her hand or finger on the poisoned spindle of a spinning wheel and falls into a deep sleep following a wicked fairy or witch's curse. Numerous variations of the tale exist (the Brothers Grimm had one in their collection entitled Little Briar Rose), and in only some of them is the spindle actually attached to/associated with a spinning wheel.

Perhaps surprisingly, a traditional spindle does not have a sharp end that could prick a person's finger (unlike the walking wheel, often used for wool spinning). Despite this, the narrative idea persists that Sleeping Beauty or Briar Rose or Dornrosen pricks her finger on the spindle—a device which she has never seen before, as they have been banned from the kingdom in a forlorn attempt to prevent the curse of the wicked godmother-fairy.

Walt Disney included the Saxony or flax wheel in their animated film version of Perrault's tale and Rose pricks her finger on the distaff (which holds the plant fiber waiting to be spun). Whereas only a spindle is used in Tchaikovsky's ballet *The Sleeping Beauty* which is closer to the direct translation of the French "un fuseau". Spinning wheels are also integral to the plot or characterization in the Scottish folk tale *Habitrot* and the German tales *The Three Spinners*. and *The Twelve Huntsmen*

Louisa May Alcott, most famous as the author of *Little Women*, wrote a collection of short stories called *Spinning-Wheel Stories*, which were not about spinning wheels but instead meant to be read while engaging in the rather tedious act of using a spinning wheel.

Music

Classical and Symphonic

In 1814, Franz Schubert composed "Gretchen am Spinnrade", a lied for piano and voice based on a poem from Goethe's *Faust*. the piano part depicts Gretchen's restlessness as she spins on a spinning wheel while waiting by a window for her love to return.

Antonín Dvořák composed *The Golden Spinning Wheel*, a symphonic poem based on the folk ballad from *Kytice* by Karel Jaromír Erben.

Camille Saint-Saëns wrote *Le Rouet d'Omphale (Omphale's Spinning Wheel)*, symphonic poem in A major, Op. 31, a musical treatment of the classical story of Omphale and Heracles.

A favorite piano work for students is Albert Ellmenreich's *Spinnleidchen (Spinning Song)*, from his 1863 *Musikalische Genrebilder*, Op. 14. An ostinato of repeating melodic fifths represents the spinning wheel.

Folk and Ballad

The Spinning Wheel is also the title/subject of a classic Irish folk song by John Francis Waller.

A traditional Irish folk song, *Túirne Mháire*, is generally sung in praise of the spinning wheel, but was regarded by Mrs Costelloe, who collected it, as "much corrupted", and may have had a darker narrative. It is widely taught in junior schools in Ireland.

Sun Charkhe Di Mithi Mithi Kook is a Sufi song in the Punjabi language inspired by the traditional spinning wheel.

Opera

Spinning wheels also feature prominently in the Wagner opera *The Flying Dutchman*; the second act begins with local girls sitting at their wheels and singing about the act of spinning.

Art

Spinning wheels may be found as motifs in art around the world, ranging from their status as domestic/utilitarian items to their more symbolic role (such as in India, where they may have political implications).

Spinning Mule

The only surviving example of a spinning mule built by the inventor Samuel Crompton

The spinning mule is a machine used to spin cotton and other fibres in the mills of Lancashire and elsewhere. They were used extensively from the late 18th to the early 20th century. Mules were worked in pairs by a minder, with the help of two boys: the little piecer and the big or side piecer. The carriage carried up to 1,320 spindles and could be 150 feet (46 m) long, and would move forward and back a distance of 5 feet (1.5 m) four times a minute. It was invented between 1775 and 1779 by Samuel Crompton. The self-acting (automatic) mule was patented by Richard Roberts in

1825. At its peak there were 50,000,000 mule spindles in Lancashire alone. Modern versions are still in niche production and are used to spin woollen yarns from noble fibres such as cashmere, ultra-fine merino and alpaca for the knitware market.

The spinning mule spins textile fibres into yarn by an intermittent process. In the draw stroke, the roving is pulled through rollers and twisted; on the return it is wrapped onto the spindle. Its rival, the throstle frame or ring frame uses a continuous process, where the roving is drawn, twisted and wrapped in one action. The mule was the most common spinning machine from 1790 until about 1900 and was still used for fine yarns until the early 1980s. In 1890, a typical cotton mill would have over 60 mules, each with 1,320 spindles, which would operate four times a minute for 56 hours a week.

History

Before the 1770s, textile production was a cottage industry using flax and wool. Weaving was a family activity. The children and women would card the fibre, which the women would spin into yarn; the male weaver would use a frame loom to weave this into cloth. This was then tentered in the sun to bleach it. The same production system was attempted with cotton, but the demand was too high, and with the invention by John Kay of the flying shuttle, which made the loom twice as productive, more cotton yarn was being woven than the traditional spinners could supply. There were two types of spinning wheel: the Simple Wheel, which uses an intermittent process, and the more refined Saxony wheel, which drives a differential spindle and flyer with a heck (an apparatus that guides the thread to the reels) in a continuous process. These two wheels became the starting point of technological development. Businessmen such as Richard Arkwright employed inventors to find solutions that would increase the amount of yarn spun, then took out the relevant patents. The spinning jenny allowed a group of eight spindles to be operated together. It mirrored the simple wheel; the rovings were clamped and a frame moved forward stretching and thinning the roving. A wheel was rapidly turned as the frame was pushed back, and the spindles rotated, twisted the rovings into yarn and collecting it on the spindles.

The throstle and the later water frame pulled the rovings through a set of attenuating rollers, spinning at differing speeds these pulled the thread continuously and it was twisted by the heck as it was wound on the heavy spindles. Eight or sixteen of these were mounted in parallel on a static frame driven usually by a water wheel. It was ideas from these two system that inspired the spinning mule. It was the water frame that inspired the ring frame.

The increased supply of muslin inspired developments in loom design such as Edmund Cartwright's power loom. Some spinners and handloom weavers opposed the perceived threat to their livelihood: there were frame-breaking riots and, in 1811–13, the Luddite riots. The preparatory and associated tasks allowed many children to be employed until this was regulated.

The hand-operated mule was a breakthrough in yarn production and the machines were copied by Samuel Slater, who founded the cotton industry in Rhode Island. Development over the next century and a half led to an automatic mule and to finer and stronger yarn. The ring frame, originating in New England in the 1820s, was little used in Lancashire until the 1890s. It required more energy and could not produce the finest counts.

The First Mule

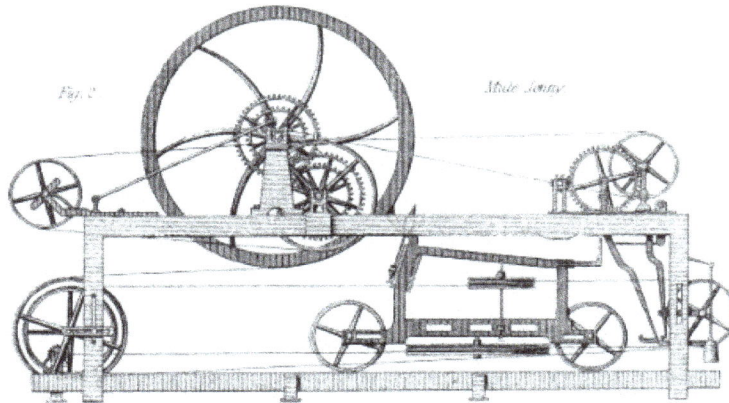

An early spinning mule: showing the gearing in the headstock

Samuel Crompton invented the spinning mule in 1779, so called because it is a hybrid of Arkwright's water frame and James Hargreaves' spinning jenny in the same way that mule is the product of crossbreeding a female horse with a male donkey. The spinning mule has a fixed frame with a creel of bobbins to hold the roving, connected through the headstock to a parallel carriage with the spindles. On the outward motion, the rovings are paid out through attenuating rollers and twisted. On the return, the roving is clamped and the spindles reversed to take up the newly spun thread.

Crompton built his mule from wood. Although he used Hargreaves' ideas of spinning multiple threads and of attenuating the roving with rollers, it was he who put the spindles on the carriage and fixed a creel of roving bobbins on the frame. Both the rollers and the outward motion of the carriage remove irregularities from the rove before it is wound on the spindle. When Arkwright's patents expired, the mule was developed by several manufacturers. Crompton's first mule had 48 spindles and could produce 1 lb of 60s thread a day. This demanded a spindle speed of 1,700 rpm, and a power input of 1/16 hp.

The mule produced strong, thin yarn, suitable for any kind of textile. It was first used to spin cotton, then other fibres.

Samuel Crompton could not afford to patent his invention. He sold the rights to David Dale and returned to weaving. Dale patented the mule and profited from it.

Improvements

Crompton's machine was largely built of wood, using bands and pulley for the driving motions. After his machine was public, he had little to do with its development. Henry Stones, a mechanic from Horwich, constructed a mule using toothed gearing and, importantly, metal rollers. Baker of Bury worked on drums, and Hargreaves used parallel scrolling to achieve smoother acceleration . and deceleration.

In 1790, William Kelly of Glasgow used a new method to assist the draw stroke. First animals and then water was used as the prime mover. Wright of Manchester moved the head stock to the centre of the machine, allowing twice as many spindles; a squaring band was added to ensure the spindles came out in a straight line He was in conversation with John Kennedy about the possibility of a self-acting mule. Kennedy, a partner in McConnell & Kennedy machine makers in Ancoats, was concerned with building ever larger mules. McConnell & Kennedy ventured into spinning when they were left with two unpaid-for mules; their firm prospered and eventually merged into the Fine Spinners & Doublers Association. In 1793, John Kennedy was addressing the problem of fine counts. With these counts, the spindles on the return traverse needed to rotate faster than on the outward traverse. He attached gears and a clutch to implement this motion.

William Eaton, in 1818, improved the winding of the thread by using two faller wires and performing a backing off at the end of the outward traverse. All these mules had been worked by the strength of the operatives. The next improvement was a fully automatic mule.

Roberts' Self-acting Mule

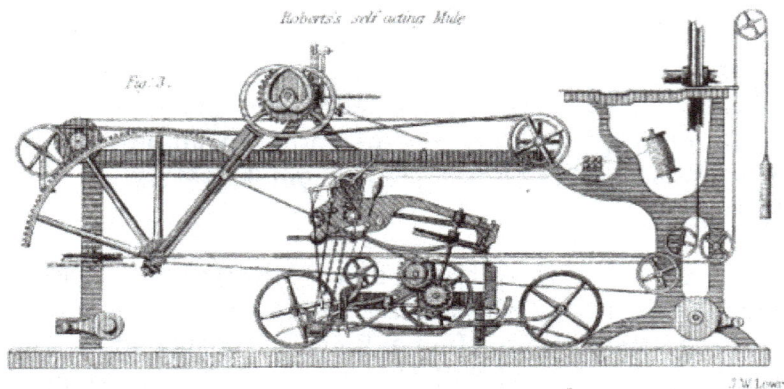

A Roberts self-acting spinning mule:1835 diagram showing the gearing in the headstock

Richard Roberts took out his first patent in 1825 and a second in 1830. The task he had set himself was to design a self-actor, a self-acting or automatic spinning mule. Roberts is also known for the Roberts Loom, which was widely adopted because of its reliability. The mule in 1820 still needed manual assistance to spin a consistent thread; a self-acting mule would need:

- A reversing mechanism that would unwind a spiral of yarn on the top of each spindle, before commencing the winding of a new stretch

- A faller wire that would ensure the yarn was wound into a predefined form such as a cop

- An appliance to vary the speed of revolution of the spindle, in accordance with the diameter of thread on that spindle

A counter faller under the thread was made to rise to take in the slack caused by backing off. This could be used with the top faller wire to guide the yarn to the correct place on the cop. These were controlled by levers and cams and an inclined plane called the shaper. The spindle speed was controlled by a drum and weighted ropes, as the headstock moved the ropes twisted the drum, which using a tooth wheel turned the spindles. None of this would have been possible using the technology of Crompton's time, fifty years earlier.

The outward traverse

The inward traverse

Notice the faller wire gear

With the invention of the self actor, the hand operated mule was increasingly referred to as a mule-jenny.

Oldham Counts

Oldham counts refers to the medium thickness cotton that was used for general purpose cloth. Roberts didn't profit from his self-acting spinning mule, but on the expiry of the patent other firms took forward the development, and the mule was adapted for the counts it spun. Initially Robert's self-actor was used for coarse counts (Oldham Counts), but the mule-jenny continued to be used for the very finest counts (Bolton counts) until the 1890s and beyond.

Bolton Counts

Bolton specialised in fine count cotton, and its mules ran more slowly to put in the extra twist. The mule jenny allowed for this gentler action but in the 20th century additional mechanisms were added to make the motion more gentle, leading to mules that used two or even three driving speeds. Fine counts needed a softer action on the winding, and relied on manual adjustment to wind the chase or top of the perfect cop,

Woollen Mules

Spinning wool was very different: the staple was naturally twisted and easily adhered to other staples. The yarn could be bulked out by pressing in short fibres that would have been consider too short to spin if cotton. The mule could be far simpler in its construction.

Condenser Spinning

A pair of Condenser spinning mules. These have 741 spindles, being cut down from 133 feet (41 m) 1,122 spindles they used to have up until 24 September 1974, when they were retired from Elk Mill, Royton. The mule was built by Platt Brothers, of Oldham in 1927 for Elk, the last spinning mill built.

Condenser spinning or cotton waste spinning is akin to spinning wool, and the mules are similar. Helmshore Mills was a cotton waste mule spinning mill.

Current Usage

Mules are still in use for spinning woolen and alpaca, and being produced across the world. In Italy for example by Bigagli and Cormatex

Operation of a Mule

Taylor, Lang & Co selfactor mule headstock

Mule spindles rest on a carriage that travels on a track a distance of 60 inches (1.5 m), while drawing out and spinning the yarn. On the return trip, known as putting up, as the carriage moves back to its original position, the newly spun yarn is wound onto the spindle in the form of a cone-shaped cop. As the mule spindle travels on its carriage, the roving which it spins is fed to it through rollers geared to revolve at different speeds to draw out the yarn.

Marsden in 1885 described the processes of setting up and operating a mule. Here is his description, edited slightly.

The creel holds bobbins containing rovings. The rovings are passed through small guide-wires, and between the three pairs of drawing-rollers.

- The first pair takes hold of the roving, to draw the roving or sliver from the bobbin, and deliver it to the next pair.

- The motion of the middle pair is slightly quicker than the first, but only sufficiently so to keep the roving uniformly tense

- The front pair, running much more quickly, draws out (attenuates) the roving so it is equal throughout.

Connection is then established between the attenuated rovings and the spindles. When the latter are bare, as in a new mule, the spindle-driving motion is put into gear, and the attendants wind upon each spindle a short length of yarn from a cop held in the hand. The drawing-roller motion is placed in gear, and the rollers soon present lengths of attenuated roving. These are attached to the threads on the spindles, by simply placing the threads in contact with the un-twisted roving. The different parts of the machine are next simultaneously started, when the whole works in harmony together.

The back rollers pull the sliver from the bobbins, and passing it to the succeeding pairs, whose differential speeds attenuate it to the required degree of fineness. As it is delivered in front, the spindles, revolving at a rate of 6,000–9,000 rpm twist the hitherto loose fibres together, thus forming a thread.

Whilst this is going on, the spindle carriage is being drawn away from the rollers, at a pace very slightly exceeding the rate at which the roving is coming forth. This is called the gain of the carriage, its purpose being to eliminate all irregularities in the fineness of the thread. Should a thick place in the roving come through the rollers, it would resist the efforts of the spindle to twist it; and, if passed in this condition, it would seriously deteriorate the quality of the yarn, and impede subsequent operations. As, however, the twist, spreading itself over the level thread, gives firmness to this portion, the thick and untwisted part yields to the draught of the spindle, and, as it approaches the tenuity of the remainder, it receives the twist it had hitherto refused to take. The carriage, which is borne upon wheels, continues its outward progress, until it reaches the extremity of its traverse, which is 63 inches (160 cm) from the roller beam. The revolution of the spindles cease, the drawing rollers stop.

Backing-off commences. This process is the unwinding of the several turns of the yarn, extending from the top of the cop in process of formation to the summit of the spindle. As this proceeds, the faller- wire, which is placed over and guides the threads upon the cop, is depressed ; the counter-faller at the same time rising, the slack unwound from the spindles is taken up, and the threads are prevented from running into snarls. Backing-off is completed.

The carriage commences to run inwards; that is, towards the rollerbeam. This is called putting up. The spindles wind on the yarn at a uniform rate. The speed of revolution of the spindle must vary, as the faller is guiding the thread upon the larger or smaller diameter of the cone of the cop. Immediately the winding is finished, the depressed faller rises, the counter-faller is put down.

These movements are repeated until the cops on each spindle are perfectly formed: the ' set is completed. A stop-motion paralyzes every action of the machine, rendering it necessary to doff or strip the spindles, and to commence anew.

Doffing is performed by the piercers thrutching, that is raising, the cops partially up the spindles, whilst the carriage is out. The minder then depressing the faller, so far as to guide the threads upon the bare spindle below. A few turns are wound onto the spindle, to fix the threads to the bare spindles for a new set. The cops are removed and collected into cans or baskets, and subsequently delivered to the warehouse. The remainder of the "draw" or "stretch," as the length of spun yarn is called when the carriage is out, is then wound upon the spindles as the carriage is run up to the roller beam. Work then commences anew. The doffing took only a few minutes, the piercers would run the length of the mule gate thrutching five spindles a time, and the doffing involved lifting four cops from the spindles with the right hand and piling them on the left forearm and hand. To get a firm cop bottom, the minder would whip the first few layers of yarn. After the first few draws the minder would stop the mule at the start of an inward run and take it in slowly depressing and releasing the faller wire several times. Alternatively, a starch paste could be skillfully applied to the first few layers of yarn by the piercers – and later a small paper tube was dropped over spindle – this slowed down the doffing operation and extra payment was negotiated by the minders.

Duties of the Operatives

A pair of mules would be manned by a person called the minder and two boys called the side piecer and the little piecer. They worked barefoot in humid temperatures, the minder and the little piecer worked the minder's half of the mule. The minder would make minor adjustments to his mules to the extent that each mule worked differently. They were specialists in spinning, and were only answerable to the gaffer and under-gaffer who were in charge of the floor and with it the quantity and quality of the yarn that was produced. Bobbins of rovings came from the carder in the blowing room delivered by a bobbin carrier who was part of the carder's staff, and yarn was hoisted down to the warehouse by the warehouseman's staff. Delineation of jobs was rigid and communication would be through the means of coloured slips of paper written on in indelible pencil.

Mule-spinning Room

Creeling involved replacing the rovings bobbins in a section of the mule without stopping the mule. On very coarse counts a bobbin lasted two days but on fine count it could last for 3 weeks. To creel, the creeler stood behind the mule, he placed new bobbins on the shelf above the creel. As the bobbin ran empty he would pick it off its skewer in the creel unreeling 30 cm or so of roving, and drop it into a skip. With his left hand, he would place on the new bobbin onto the skewer from above and with his right hand twist in the new roving into the tail of the last.

Piecing involved repairing sporadic yarn breakages. At the rollers, the broken yarn would be caught on the underclearer (or fluker rod on Bolton mules), while at the spindle it would knot itself into a whorl on the spindle tip. If the break happened on the winding stroke the spindle might have to be stopped while the thread was found. The number of yarn breakages was dependent on the quality of the roving, and quality cotton led to fewer breakages. Typical 1,200 spindle mules of the 1920s would experience 5 to 6 breakages a minute. The two piecers would thus need to repair the thread within 15 to 20 seconds while the mule was in motion but once they had the thread it took under three seconds. The repair actually involved a slight rolling of the forefinger against the thumb.

Doffing has Already Been Described.

Cleaning was important and until a formal ritual had been devised it was a dangerous operation. The vibration in a mule threw a lot of short fibres (or fly) into the air. It tended to accumulate on the carriage behind the spindles and in the region of the drafting rollers. Piking the stick meant placing the hand though the yarnsheet, and unclipping two sticks of underclearer rollers from beneath the drafting rollers, drawing them through the 1 ¼ in gap between two ends, stripping them of fly and replacing them on the next inward run. Cleaning the carriage top was far more dangerous. The minder would stop the mule on the outward run, and raise his hands above his head. The piecers would enter under the yarn sheet with a scavenger cloth on the carriage spindle rail and a brush on the roller beam, and run bent double the entire length of the mule, avoiding the rails and draw bands, and not letting themselves touch the yarn sheet. When they had finished they would run to agreed positions of safety where the minder could see both of them, and the minder would unclip the stang and start the mule. Before this ritual was devised, boys had been crushed. The mule was 130 feet (40 m) long, the minder's eyesight might not have been good, the air in the mill was clouded with fly and another minder's boys might have been mistaken for his. The ritual became encoded in law.

Key Components

A Mule Jenny 1892

- Drawing rollers
- Faller and counter faller
- Quadrant

Terminology

Social and Economic

The spinning inventions were significant in enabling a great expansion to occur in the production of textiles, particularly cotton ones. Cotton and iron were leading sectors in the Industrial Revolution. Both industries underwent a great expansion at about the same time, which can be used to identify the start of the Industrial Revolution.

Mules operating in a Cotton mill.

The 1790 mule was operated by brute force: the spinner drawing and pushing the frame while attending to each spindle. Home spinning was the occupation of women and girls, but the strength needed to operate a mule caused it to be the activity of men. Hand loom weaving, however, had been a man's occupation but in the mill it could and was done by girls and women. Spinners were the bare-foot aristocrats of the factory system.

Mule spinners were the leaders in unionism within the cotton industry; the pressure to develop the self-actor or self-acting mule was partly to open the trade to women. It was in 1870 that the first national union was formed.

The wool industry was divided into woollen and worsted. It lagged behind cotton in adopting new technology. Worsted tended to adopt Arkwright water frames which could be operated by young girls, and woollen adopted the mule.

Mule-spinners' Cancer

About 1900 there was a high incidence of scrotal cancer detected in former mule spinners. It was limited to cotton mule spinners and did not affect woollen or condenser mule spinners. The cause was attributed to the blend of vegetable and mineral oils used to lubricate the spindles. The spindles when running threw out a mist of oil at crotch height, that was captured by the clothing of anyone piecing an end. In the 1920s much attention was given to this problem. Mules had used this mixture since the 1880s, and cotton mules ran faster and hotter than the other mules, and needed more frequent oiling. The solution was to make it a statutory requirement to only use vegetable oil or white mineral oils, which were believed to be non-carcinogens. By then cotton mules had been superseded by the ring frame and the industry was contracting, so it was never established whether these measures were effective.

Ring Spinning

A ring spinning machine in the 1920s

Ring spinning is a method of spinning fibres, such as cotton, flax or wool, to make a yarn. The ring frame developed from the throstle frame, which in its turn was a descendant of Arkwright's water frame. Ring spinning is a continuous process, unlike mule spinning which uses an intermittent action. In ring spinning, the roving is first attenuated by using drawing rollers, then spun and wound around a rotating spindle which in its turn is contained within an independently rotating ring flyer. Traditionally ring frames could only be used for the coarser counts, but they could be attended by semi-skilled labour.

History

Early Machines

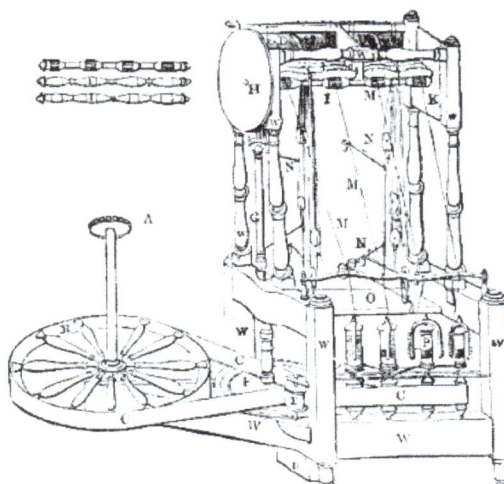

Arkwright's spinning frame

- The Saxony wheel was a double band treadle spinning wheel. The spindle rotated faster than the traveller in a ratio of 8:6, drawing was done by the spinners fingers.

- The water frame was developed and patented by Arkwright in the 1770s. The roving was attenuated (stretched) by draughting rollers and twisted by winding it onto a spindle. It was heavy large-scale machine that needed to be driven by power, which in the late 18th century meant by a water wheel. Cotton mills were designed for the purpose by Arkwright, Jedediah Strutt and others along the River Derwent in Derbyshire. Water frames could only spin weft.

- The throstle frame was a descendant of the water frame. It used the same principles, was better engineered and driven by steam. In 1828 the Danforth throstle frame was invented in the United States. The heavy flyer caused the spindle to vibrate, and the yarn snarled every time the frame was stopped. Not a success.

- The Ring frame is credited to John Thorp in Rhode Island in 1828/9 and developed by Mr. Jencks of Pawtucket, Rhode Island, who Richard Marsden names as the inventor.

Developments in the United States

Machine shops experimented with ring frames and components in the 1830s. The success of the ring frame, however, was dependent on the market it served and it was not until industry leaders like Whitin Machine Works in the 1840s and the Lowell Machine Shop in the 1850s began to manufacture ring frames that the technology started to take hold.

At the time of the American Civil War, the American industry boasted 1,091 mills with 5,200,000 spindles processing 800,000 bales of cotton. The largest mill, Naumkeag Steam Cotton Co. in Salem, Mass.had 65,584 spindles. The average mill housed only 5,000 to 12,000 spindles, with mule spindles out-numbering ring spindles two-to-one.

After the war, mill building started in the south, it was seen as a way of providing employment. Almost exclusively these mills used ring technology to produce coarse counts, and the New England mills moved into fine counts.

Jacob Sawyer vastly improved spindle for the ring frame in 1871, taking the speed from 5000rpm to 7500rpm and reducing the power needed, formerly 100 spindles would need 1 hp but now 125 could be driven. This also led to production of fine yarns. During the next ten years, the Draper Corporation protected its patent through the courts. One infringee was Jenks, who was marketing a spindle known after its designer, Rabbeth. When they lost the case, Mssrs. Fales and Jenks, revealed a new patent free spindle also designed by Rabbeth, and also named the Rabbeth spindle.

The Rabbeth spindle was self-lubricating and capable of running without vibration at over 7500 rpm. The Draper Co. bought the patent and expanded the Sawyer Spindle Co. to manufacture it. They licensed it to Fales & Jenks Machine Co., the Hopedale Machine Co., and later, other machine builders. From 1883 to 1890 this was the standard spindle, and William Draper spent much of his time in court defending this patent.

Adoption in Europe

The new method was compared with the self-acting spinning mule which was developed by Richard Roberts using the more advanced engineering techniques in Manchester. The ring frame was reliable for coarser counts while Lancashire spun fine counts as well. The ring frame was heavier, requiring structural alteration in the mills and needed more power. These were not problems in the antebellum cotton industry in New England. It fulfilled New England's difficulty in finding skilled spinners: skilled spinners were plentiful in Lancashire. In the main the requirements on the two continents were different, and the ring frame was not the method of choice for Europe at that moment.

Brooks and Doxey Ring Spinning Frame about 1890

Mr Samuel Brooks of Brooks & Doxey Manchester was convinced of the viability of the method. After a fact-finding tour to the States by his agent Blakey, he started to work on improving the frame. It was still too primitive to compete with the highly developed mule frames, let alone supersede them. He first started on improving the doubling frame, constructing the necessary tooling needed to improve the precision of manufacture. This was profitable and machines offering 180,000 spindle were purchased by a sewing thread manufacturer.

Brooks and other manufacturers now worked on improving the spinning frame. The principal cause for concern was the design of the Booth-Sawyer spindle. The bobbin did not fit tightly on the spindle and vibrated wildly at higher speeds. Howard & Bullough of Accrington used the Rabbath spindle, which solved these problems. Another problem was ballooning, where the thread built up in an uneven manner. This was addressed by Furniss and Young of Mellor Bottom Mill, Mellor by attaching an open ring to the traverse or ring rail. This device controlled the thread, and consequently a lighter traveller could be made which could operate at higher speeds. Another problem was the accumulation of fluff on the traveller breaking the thread - this was eliminated by a device called a traveller cleaner.

A major time constraint was doffing, or changing the spindles. Three hundred or more spindles had to be removed, and replaced. The machine had to be stopped while the doffers, who were often very young boys, did this task. The frame was idle until it was completed. A mechanical doffer system reduced the doffing time to 30–35 seconds.

Rings and Mules

The ring frame was extensively used in the United States, where coarser counts were manufactured. Many of frame manufacturers were US affiliates of the Lancashire firms, such as Howard & Bullough and Tweedales and Smalley. They were constantly trying to improve the speed and quality of their product. The US market was relatively small, the total number of spindles in the entire United States was barely more than the number of spindles in one Lancashire town, Oldham. When production in Lancashire peaked in 1926 Oldham had 17.669 million spindles and the UK had 58.206 million.

Technologically mules were more versatile. The mules were more easily changed to spin the larger variety of qualities of cotton then found in Lancashire. While Lancashire concentrated on "Fines" for export, it also spun a wider range, including the very coarse wastes. The existence of the Liverpool cotton exchange meant that mill owners had access to a wider selection of staples.

The wage cost per spindle is higher for ring spinning. In the states, where cotton staple was cheap, the additional labour costs of running mules could be absorbed, but Lancashire had to pay shipment costs. The critical factor was the availability of labour, when skilled labour was scarce then the ring became advantageous. This had always been so in New England, and when it became so in Lancashire, ring frames started to be adopted.

The first known mill in Lancashire dedicated to ring spinning was built in Milnrow for the New Ladyhouse Cotton Spinning Company (registered 26 April 1877). A cluster of smaller mills developed which between 1884 and 1914 out performed the ring mills of Oldham. After 1926, the Lancashire industry went into sharp decline, the Indian export market was lost, Japan was self-sufficient. Textile firms united to reduce capacity rather than to add to it. It wasn't until the late 1940s that some replacement spindles started to be ordered and ring frames became dominant. Debate still continues in academic papers on whether the Lancashire entrepreneurs made the right purchases decisions in the 1890s. The engine house and steam engine of the Ellenroad Ring Mill are preserved.

New Technologies

- The search for faster and more reliable ring spinning techniques continues. In 2005, a PhD paper was written at Auburn University, Alabama on using magnetic levitation to reduce friction, a techniques known as Magnetic ring spinning.

- Open end spinning was developed in Czechoslovakia in the years preceding 1967. It was far faster than ring spinning, and did away with many preparatory processes. Put simply, the thread was ejected spinning from a nozzle, and on exiting hooked onto other loose fibres in the chamber behind. It was first introduced into the United Kingdom at the Maple Mill, Oldham.

How it Works

A ring frame was constructed from cast iron, and later pressed steel. On each side of the frame are the spindles, above them are draughting (drafting) rollers and on top is a creel loaded with bobbins of roving. The roving (unspun thread) passes downwards from the bobbins to the draughting rollers. Here the back roller steadied the incoming thread, while the front rollers rotated faster, pulling the roving out and making the fibres more parallel. The rollers are individually adjustable, originally by mean of levers and weights. The attenuated roving now passes through a thread guide that is adjusted to be centred above the spindle. Thread guides are on a thread rail which allows them to be hinged out of the way for doffing or piecing a broken thread. The attenuated roving passes down to the spindle assembly, where it is threaded though a small D ring called the traveller. The traveller moves along the ring. It is this that gives the ring frame its name. From here the thread is attached to the existing thread on the spindle.

Modern ring spinning frame1 Drafting rollers2 Spindle3 Attenuated roving4 Thread guides5 Anti-ballooning ring6 Traveller7 Rings8 Thread on bobbin

The traveller, and the spindle share the same axis but rotate at different speeds. The spindle is driven and the traveller drags behind thus distributing the rotation between winding up on the spindle and twist into the yarn. The bobbin is fixed on the spindle. In a ring frames, the different speed was achieved by drag caused by air resistance and friction (lubrication of the contact surface between the traveller and the ring was a necessity). Spindles could rotate at speeds up to 25,000 rpm, this spins the yarn. The up and down ring rail motion guides the thread onto the bobbin into the shape required: i.e. a cop. The lifting must be adjusted for different yarn counts.

Doffing is a separate process. An attendant (or robot in an automated system) winds down the ring rails to the bottom. The machine stops. The thread guides are hinged up. The completed bobbin coils (yarn packages) are removed from the spindles. The new bobbin tube is placed on the spindle trapping the thread between it and the cup in the wharf of the spindle, the thread guides are lowered and the machine restarted. Now all the processes are done automatically. The yarn is taken to a cone winder. Currently, machines are manufactured by Rieter (Switzerland), ToyoTa (Japan), Zinser, Suessen, (Germany) and Marzoli (Italy). The Rieter compact K45 system, has 1632 spindles, while the ToyoTa has a machine with 1824 spindles. All require controlled atmospheric conditions.

Open-end Spinning

Open-end spinning is a technology for creating yarn without using a spindle. It was invented and developed in Czechoslovakia in Výzkumný ústav bavlnářský / Cotton Research Institute in Ústí nad Orlicí in 1963.

Method

It is also known as break spinning or rotor spinning. The principle behind open-end spinning is similar to that of a clothes dryer spinning full of sheets. If you could open the door and pull out a sheet, it would spin together as you pulled it out. Sliver from the card goes into the rotor, is spun into yarn and comes out, wrapped up on a bobbin, all ready to go to the next step. There is no roving stage or re-packaging on an auto-coner. This system is much less labour-intensive and faster than ring spinning with rotor speeds up to 140,000 rpm. The Rotor design is the key to the operation of the open-ended spinners. Each type of fibre may require a different rotor design for optimal product quality and processing speed.

The first open-end machines in the United Kingdom were placed, under great secrecy, by Courtaulds into Maple Mill, Oldham in 1967.

One disadvantage of open-end spinning is that it is limited to coarser counts, another is the structure of the yarn itself with fibres less in parallel compared to ring-spun yarns, for example, consequently cloth made from open-end yarn has a "fuzzier" feel and poorer wear resistance.

History

The global demand for spun fibre is huge. Converting raw fibre to yarn is a complicated process. Many manufacturers compete to provide the spinning machines that are essential to meeting the demand by delivering increases in spinning productivity and additional improvements in yarn quality. Over the past three centuries spinning technology has been continuously improved through thousands of minor innovations, and occasional major advances that have collectively increased the quality and lowered the cost of producing yarn dramatically.

Major technology advances have included:

- Hand spinning

- Mule spinning

- Ring spinning

- Rotor spinning

- Dref Friction Spinning

- Open-end spinning

Development stages of open-end spinning	
1937	Berthelsen developed a relatively perfect open end.
1965	Czech KS200 rotor spinning machine was introduced at 30000 rotor rpm.
1967	Improved BD200 with G5/1 Rieter were presented with first mill of OE coming under production.
1971–1975	There was a considerable increase in machine manufacturer and newer and improved version of machines were launched with increased speed at 100000 rpm.
1975	Also witnessed first automated machine from Suessen equipped with Spincat and Cleancat which opened up the industrial rotor spinning breakthrough.
1977	Witnessed Schlafhorst with Autocoro machines, which made a mark in open-end market.

The number of manufacturers who can successfully compete has been reduced, as the technical complexity of the spinning machines has increased. However, there are many competent companies serving the global market for spinning machines who continue to pursue innovative ways to increase spinning productivity and yarn quality.

Characteristics

A good open-end machine should have:

- Higher productivity

  This is a major criterion, as productivity reduces the cost of manufacturing. The O.E. machines that are now in market boasts of many a basic needs like, longer length of machine, higher speeds, able to process coarser hank, fewer changes for count, easy access to parts (less downtime for cleaning), longer production time between cleaning schedules, computerized controls for less power consumption and lower downtime and complete report generation giving leads to problem area are some points to discuss.

- High-content sliver cans (up to 18")

  In early days large machines were equipped with less distance between rotors (gauge of machine). This led to creeling of very small cans, which required frequent can changes. Each can change requires a break in the yarn. All major manufacturers currently allow cans up to 18" diameter leading to less breakage, less joining of yarn, hence better quality and higher productivity. Originally round cans were used. Rectangular cans are used because they double sliver capacity in the same sliver can footprint.

- Larger packages of yarn (4 to 5 kg)

  The final package size has continued to increase. The final package size is important because it reduces tube change frequency and thus reduces idle time for creeling. Current yarn packages typically weigh 4–5 kg. The Savio Super Spinner 3000 currently has the largest package size at 6 kg.

- Less power consumption

  Using individual motors and electronic controls for each of the various drives of the machine maximizes energy efficiency and minimizes downtime.

- Automation

  All spinning machines, whether ring or open-end, need yarn joining to repair breaks or start new sliver cans. Joining the yarn has historically been a labor-intensive activity and a source of quality defects. Autopiecing units are robots that automate this process. Market leaders like Schlafhorst, Rieter, Savio, have machines that incorporate good quality autopiecers and autodoffing. This automation leads to less material handling costs and helps improve quality of the final product.

- Flexibility of spinning components

Many vendors are offering machines that can be programmed to produce many different types of yarns. The ability to rapidly change production results in the flexibility to serve multiple markets. A contemporary spinning mill should be able to produce a range of products: denim, knitting, towels, structured fabrics, construction fabrics, and various other products like core spun, multi count, etc.

- Handling count range.

Machines need to be easily programmed to spin yarns from 4sNe to 60sNe. This ability allows a single machine to produce yarns that cater to many different end-user requirements.

Advantages

- Disappearance of simplex frame.
- Under certain circumstances, elimination of the second passage draw frame.
- In some cases, with the use of auto-leveller at the cards, elimination of even the draw frame passage.
- Bigger supply of cans to open-end and bigger packages to weaving.
- Elimination of winding.
- Less labor and power cost per kilogram of yarn.
- Higher productivity almost 7 times in the case of 10s and high efficiency.
- Fully automated mill a reality.

Disadvantages

- Restricted only coarse counts.
- High capital cost.
- Usage restricted in case yarn is weak.
- Yarn realization in the case of waste mixing will be poor, resulting in increased mixing cost.
- Wear and tear of rotors, combing rollers, and navels are very high when highly trash content mixing is used resulting in heavy replacement cost.
- In case reeling is done additional reeling cost is involved resulting in higher manufacturing cost.

Products

- Linen / Flax yarns
- Cotton Yarns

- Polyester Cotton Blended yarn

- Tencel 100%

- Polyester 100%

- Polyester / Cotton / Linen / Viscose Multi blend

- Dyed yarn (and fibre)

- Acrylic

- Recycle Polyester 100% and different Blends

Dref Friction Spinning

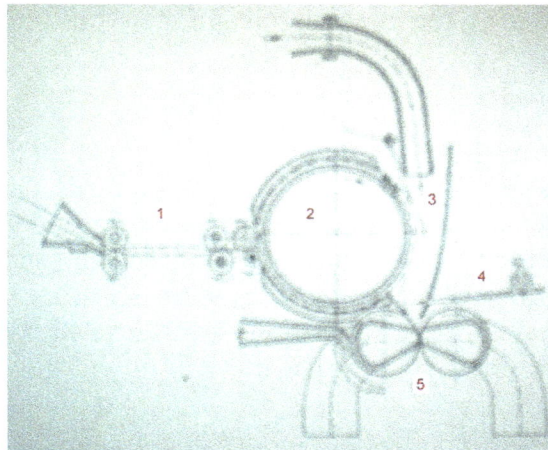

Friction spinning: air is used to propel the sliver of fibres (1) to a carding drum (2) where they drop (3) between two perforated drums (5) that integrate and twist the yarn

Friction Spinning or Dref Spinning is a textile technology that suitable for spinning coarse counts of yarns and technical core-wrapped yarns. Dref yarns are bulky, with low tensile strength making them suitable for blankets and mop yarns, they can be spun from asbestos, carbon fibres and make filters was water systems. Yarns such as Rayon and Kevlar can be spun using this method. The technology was developed around 1975 by Dr. Ernst Fehrer.

Spinning Technologies

There are three current technologies used today for spinning fibres:

1. Roving spinning which uses the legacy ring -spinning technology of the twentieth century,

2. Open end, Rotor or Break Spinning used for high quality threads

3. Dref friction spinning for other yarns.

Friction spinning is the fastest of all these techniques though the yarn is irregular and bulkier, making it suitable only for some applications.

Yarn Formation in Friction Spinning System

The mechanism of yarn formation consists of three distinct operations: feeding of fibres, fibre integration and twist insertion.

Feeding

The individual fibres are transported by air currents and deposited in the spinning zone. The mode of fibre feed has a particular effect on fibre extent and fibre configuration in yarn and on its properties. There are two methods of fibre feed:

- Direct feed

The fibres are fed directly onto the rotating fibre mass that outer part of the yarn tail.

- Indirect feed.

The fibres are first accumulated on the in-going roll and then transferred to the yarn tail.

Fibres Integration

The fibres assembles through a feed tube onto a yarn core/tail within the shear field, is provided by two rotating spinning drums and the yarn core is in between them. The shear causes sheath fibres to wrap around the yarn core. The fibre orientation is highly dependent on the decelerating fibres arriving at the assembly point through the turbulent flow. The fibres in the friction drum have two probable methods for integration of incoming fibres to the sheath. One method, the fibre assembles completely on to perforated drum before their transfer to the rotating sheath. In the other method, fibres are laid directly on to rotating sheath.

Twist Insertion

There has been much research on the twisting process in friction spinning. In friction spinning, the fibres are applied twist with more or less one at a time without cyclic differentials in tension in the twisting zone. Therefore, fibre migration may not take place in friction spun yarns. The mechanism of twist insertion for core type friction spinning and open end friction spinning are different,which are described below.

Twist Insertion in Core-type Friction Spinning

In core type friction spinning, the core, made of a filament or a bundle of staple fibres, is false twisted by the spinning drum. The sheath fibres are deposited on the false twisted core surface and are wrapped helically over the core with varying helix angles. It is believed that the false twist in the core gets removed once the yarn is emerged from the spinning drums, so that this yarn has a virtually twist-less core. However, it is quite possible for some amount of false twist to remain in the fact that the sheath entraps it during yarn formation in the spinning zone.

Twist Insertion in Open End Type Friction Spinning

In open end type friction spinning the fibres in the yarn are integrated as a stacked cone. The fibres in the surface of the yarn found more compact and good packing density than the axial fibres in the yarn.

The yarn tail can be considered as a loosely constructed conical mass of fibres, formed at the nip of the spinning drums. It is of very porous and lofty structure.The fibres rotating at very high speed.

History

Dr. Ernst Fehrer (1919-2001) invented and patented the DREF friction spinning process in 1973. He had begun work on the development of this alternative to mule, ring and rotor open end spinning with the objective of surmounting the physico-mechanical limits on capacity in yarn engineering, enhancing the production speeds.The system was name usinng letters from his honorific and name. His company Dr. Ernst Fehrer AG, Textilmaschinenfabrik, was based in Linz-Leonding, Austria. He died in December 2000 at age 81 having produced more than 1000 patents.

Fehrer began his career in research, development and inventing at age 14 writing his first patent at 18. He developed a high-speed needle loom with counterbalancing technology as well as the "DREF"system. In 1988, Fehrer received the TAPPI Nonwovens Division Award for his contributions to nonwovens manufacturing technology, and in 1994 Fehrer received Textile World's first Lifetime Achievement Award.

Development

The Dref I was in development in 1975; a three-head machine, and in 1977 the first DREF 2 for the coarse yarn count range came onto the market. In view of its success, Dr. Fehrer then created the DREF 3, which was designed for the medium yarn count range and made its debut at the ITMA '79 in Hanover, before entering serial production in 1981.

New generations of the DREF 2 followed in 1986 and 1994 and the DREF 3/96 was launched at the ITMA in Milan. The 1999 ITMA in Paris witnessed the arrival of the DREF 2000, the first of which was sold prior to the fair. Full production of the DREF 2000 commenced in the autumn of 1999 in co-ordination with presentations at the ATME, USA and the SIMAT in Argentina. In 2001, the DREF 2000 also went on display in Asia at the ITMA Singapore and in Central America at the EXINTEX, Mexico.

Fehrer entered co-operations with professional textile companies to develop the technology; Rieter AG in Switzerland and Oerlikon Schlafhorst in Germany. With this co-operation the last machine developed by DREF was the DREF 3000, which was available for testing in the new facility in Linz, Austria in 2001. Saurer AG purchased Fehrer AG in 2005. DREFCORP, along with all its associated patents and intellectual property was purchased in 2007 by Nordin Technologies – a Malaysian company – that continues to develop and manufacture DREF 2000 and DREF 3000 machines as well as continuing to serve the international market with parts for the original Fehrer Dref II, Dref III, Dref 2000 and Dref 3000 friction spinning machines.

Yarn Properties

Friction spun yarns DREF yarns have bulky appearance (100-140% bulkier than the ring spun yarns). The twist is not uniform and found with loopy yarn surface. Friction spun yarns with a high %age of core have a high stiffness. Friction spun yarns are usually weak as compared to other yarns. The yarns possess only 60% of the tenacity of ring-spun yarns and about 90% of rotor spun-yarns. The increased twist and wrapping of the sheath over the core improve the cohesion between the core and sheath and within the sheath.

The breaking elongation ring, rotor and friction spun yarns have been found to be equal. Better relative tenacity efficiency is achieved during processing of cotton on rotor and friction spinning as compared to ring spinning system.

Depending on the type of fibre, the differences in strength of these yarns differ in magnitude. It has been reported that 100% polyester yarns, this strength deficiency is 32% whereas for 100% viscose yarns, it ranges from 0-25%. On the other hand, in polyester-cotton blend, DREF yarns perform better than their ring-spun counterparts. A 70/30% blend yarn has been demonstrated to be superior in strength by 25%. The breaking strength of ring yarns to be maximum followed by the rotor yarn and then 50/50 core-sheath DREF-3 yarn.

DREF yarns have been seen to be inferior in terms of unevenness, imperfections, strength variability and hairiness. DREF yarns occupy an intermediate position between ring-spun and rotor spun yarns as far as short hairs and total hairiness s concerned. For hairs longer than 3mm, the friction spun yarns are more hairy than the ring spun yarns. Rotor spun yarns show the least value in both the values. DREF yarns are most irregular in terms of twist and linear density while ring spun yarns are most even.

Textile technologists have studied the frictional behaviour of ring, rotor, friction spun yarns of 59 and 98.4 Tex spun from cotton, polyester, viscose fibres, with varying levels of twist. The yarn to yarn and yarn to guide roller friction was measured at different sliding speeds and tension ratios. However, for polyester fibres, the rotor spun yarn showed highest friction, followed by friction and ring spun yarns.

Advantages of Friction Spinning System

The forming yarn rotates at high speed compare to other rotating elements. It can spin yarn at very high twist insertion rates (ie.300,000 twist/min). The yarn tension is practically independent of speed and hence very high production rates (up to 300 m/min) can be attainable. The yarns are bulkier than rotor yarns.

The DREF II yarns are used in many applications. Blankets for the home application range, hotels and military uses etc. DREF fancy yarns used for the interior decoration, wall coverings, draperies and filler yarn.

References

- Smith, C. Wayne; Cothren, J. Tom (1999). Cotton: Origin, History, Technology, and Production. 4. John Wiley & Sons. pp. viii. ISBN 978-0471180456.

- Cotton: Origin, History, Technology, and Production By C. Wayne Smith, Joe Tom Cothren. Page viii. Published 1999. John Wiley and Sons. Technology & Industrial Arts. 864 pages. ISBN 0-471-18045-9

- Carl A Lawrence (2010) Advances in Yarn Spinning Technology pp. 261–273, Woodhead Publishing, Oxford

ISBN 978-1-84569-444-9

- Landes, David. S. (1969). The Unbound Prometheus: Technological Change and Industrial Development in Western Europe from 1750 to the Present. Cambridge, New York: Press Syndicate of the University of Cambridge. p. 138. ISBN 0-521-09418-6.

- "Hindoo Spinning-Wheel" (PDF). The Wesleyan Juvenile Offering: A Miscellany of Missionary Information for Young Persons. WesD iionary Society. IX: 108. September 1852. Retrieved 24 February 2016.

- "Gretchen am Spinnrade ("Meine Ruh'..."), song for voice & piano, D. 118 (Op. 2) – Franz Schubert". AllMusic. Retrieved 8 August 2014.

- Image of a spinning wheel in: Al-Hariri, Al-Maqamat (les Séances). Painted by Yahya ibn Mahmud al-Wasiti, Baghdad, 1237 See: Spinning, History & Gallery lhresources.wordpress.com (retrieved March 4, 2013)

- Reel, James. "Le Rouet d'Omphale, symphonic poem in A major, Op. 31 – Camille Saint-Saëns". AllMusic. Retrieved 2012-04-05.

- "Cormatex". Modern automatic spinning mules, bale breakers and carding machines used for woolen and cashmere products (in Italian and English). 2012. Retrieved 13 December 2012.

5

Knitting: An Integrated Study

This chapter deals with the process of knitting and provides the reader with a comprehensive guide to the topic of knitting while detailing the different types of knitting like warp knitting, cable knitting, lace knitting and plaited knitting. The various types of knitting produce fabric of varying texture, stretch and symmetry. Textile manufacturing is best understood in confluence with the major topics listed in the following chapter.

Knitting

Knitting is a method by which yarn is manipulated to create a textile or fabric.

Multi-colored knitwork made in stockinette stitch.

Knitting creates multiple loops of yarn, called stitches, in a line or tube. Knitting has multiple active stitches on the needle at one time. Knitted fabric consists of a number of consecutive rows of interlocking loops. As each row progresses, a newly created loop is pulled through one or more loops from the prior row, placed on the gaining needle, and the loops from the prior row are then pulled off the other needle.

Knitting may be done by hand or by using a machine.

Different types of yarns (fibre type, texture, and twist), needle sizes, and stitch types may be used to achieve knitted fabrics with diverse properties (colour, texture, weight, heat retention, water resistance, and/or integrity).

Structure

Courses and Wales

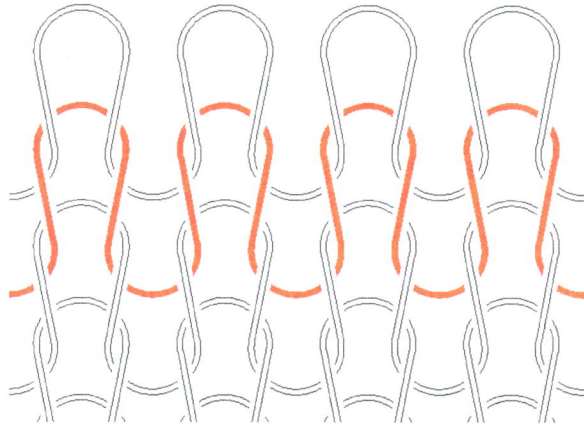

Structure of stockinette, a common knitted fabric. The meandering red path defines one *course*, the path of the yarn through the fabric. The uppermost white loops are unsecured and "active", but they secure the red loops suspended from them. In turn, the red loops secure the white loops just below them, which in turn secure the loops below them, and so on.

Alternating wales of red and yellow knit stitches. Each stitch in a wale is suspended from the one above it.

Like weaving, knitting is a technique for producing a two-dimensional fabric made from a one-dimensional yarn or thread. In weaving, threads are always straight, running parallel either lengthwise (warp threads) or crosswise (weft threads). By contrast, the yarn in knitted fabrics follows a meandering path (a *course*), forming symmetric loops (also called bights) symmetrically above and below the mean path of the yarn. These meandering loops can be easily stretched in different directions giving knit fabrics much more elasticity than woven fabrics. Depending on the yarn and knitting pattern, knitted garments can stretch as much as 500%. For this reason, knitting was initially developed for garments that must be elastic or stretch in response to the wearer's motions, such as socks and hosiery. For comparison, woven garments stretch mainly along one or other of a related pair of directions that lie roughly diagonally between the warp and the weft, while contracting in the other direction of the pair (stretching and contracting with the *bias*), and are not very elastic, unless they are woven from stretchable material such as spandex. Knitted garments are often more form-fitting than woven garments, since their elasticity allows them to contour to the body's outline more closely; by contrast, curvature is introduced into most woven garments only with sewn darts, flares, gussets and gores, the seams of which lower the elasticity of the woven fabric still further. Extra curvature can be introduced into knitted garments without seams, as in the heel of a sock; the effect of darts, flares, etc. can be obtained with short rows or by increasing or decreasing the number of stitches. Thread used in weaving is usually much finer than the yarn used in knitting, which can give the knitted fabric more bulk and less drape than a woven fabric.

If they are not secured, the loops of a knitted course will come undone when their yarn is pulled; this is known as *ripping out, unravelling* knitting, or humorously, *frogging* (because you 'rip it', this sounds like a frog croaking: 'rib-bit'). To secure a stitch, at least one new loop is passed through it. Although the new stitch is itself unsecured ("active" or "live"), it secures the stitch(es) suspended from it. A sequence of stitches in which each stitch is suspended from the next is called a *wale*. To secure the initial stitches of a knitted fabric, a method for casting on is used; to secure the final stitches in a wale, one uses a method of binding/casting off. During knitting, the active stitches are secured mechanically, either from individual hooks (in knitting machines) or from a knitting needle or frame in hand-knitting.

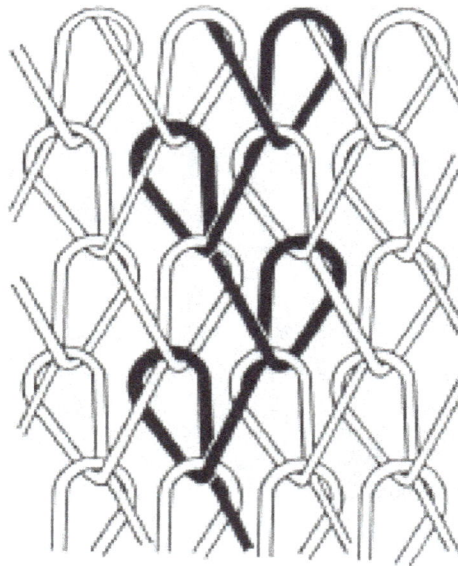

Basic pattern of warp knitting. Parallel yarns zigzag lengthwise along the fabric, each loop securing a loop of an adjacent strand from the previous row.

Weft and Warp Knitting

There are two major varieties of knitting: weft knitting and warp knitting. In the more common *weft knitting*, the wales are perpendicular to the course of the yarn. In warp knitting, the wales and courses run roughly parallel. In weft knitting, the entire fabric may be produced from a single yarn, by adding stitches to each wale in turn, moving across the fabric as in a raster scan. By contrast, in warp knitting, one yarn is required for every wale. Since a typical piece of knitted fabric may have hundreds of wales, warp knitting is typically done by machine, whereas weft knitting is done by both hand and machine. Warp-knitted fabrics such as tricot and milanese are resistant to runs, and are commonly used in lingerie.

A modern knitting machine in the process of weft knitting

Weft-knit fabrics may also be knit with multiple yarns, usually to produce interesting color patterns. The two most common approaches are intarsia and stranded colorwork. In intarsia, the yarns are used in well-segregated regions, e.g., a red apple on a field of green; in that case, the yarns are kept on separate spools and only one is knitted at any time. In the more complex stranded approach, two or more yarns alternate repeatedly within one row and all the yarns must be carried along the row, as seen in Fair Isle sweaters. Double knitting can produce two separate knitted fabrics simultaneously (e.g., two socks). However, the two fabrics are usually integrated into one, giving it great warmth and excellent drape.

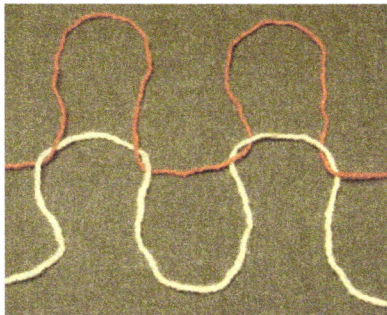

In the knit stitch on the left, the next (red) loop passes through the previous (yellow) loop from *below*, whereas in the purl stitch (right), the next stitch enters from above. Thus, a knit stitch on one side of the fabric appears as a purl stitch on the other, and vice versa.

Knit and Purl Stitches

In securing the previous stitch in a wale, the next stitch can pass through the previous loop from either below or above. If the former, the stitch is denoted as a *knit stitch* or a *plain stitch*; if the latter, as a *purl stitch*. The two stitches are related in that a knit stitch seen from one side of the fabric appears as a purl stitch on the other side.

The two types of stitches have a different visual effect; the knit stitches look like "V"'s stacked vertically, whereas the purl stitches look like a wavy horizontal line across the fabric. Patterns and pictures can be created in knitted fabrics by using knit and purl stitches as "pixels"; however, such pixels are usually rectangular, rather than square, depending on the gauge/tension of the knitting.

Individual stitches, or rows of stitches, may be made taller by drawing more yarn into the new loop (an elongated stitch), which is the basis for uneven knitting: a row of tall stitches may alternate with one or more rows of short stitches for an interesting visual effect. Short and tall stitches may also alternate within a row, forming a fish-like oval pattern.

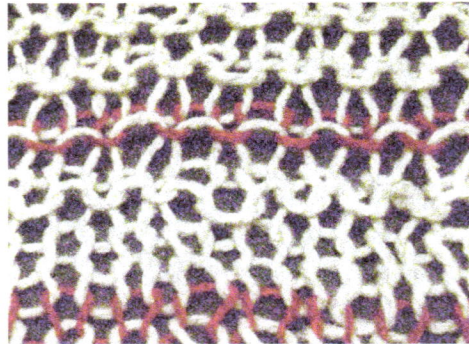

Two courses of red yarn illustrating two basic fabric types. The lower red course is knit into the white row below it and is itself knit on the next row; this produces *stockinette* stitch. The upper red course is purled into the row below and then is knit, consistent with *garter* stitch.

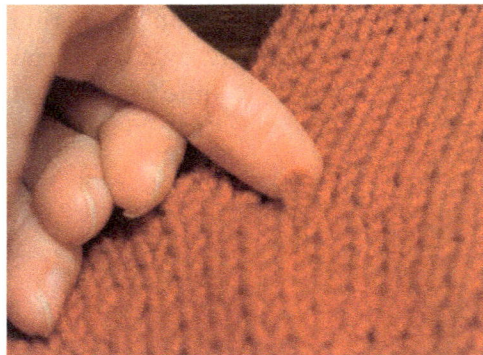

A dropped stitch, or missed stitch, is a common error that creates an extra loop to be fixed.

In the simplest knitted fabrics, all the stitches are knit or purl; this is known as a garter stitch. Alternating rows of knit stitches and purl stitches produce what is known as a stockinette pattern/stocking stitch. Vertical stripes (ribbing) are possible by having alternating wales of knit and purl stitches. For example, a common choice is 2x2 ribbing, in which two wales of knit stitches are followed by two wales of purl stitches, etc. Horizontal striping (welting) is also possible, by alternating *rows* of knit and purl stitches. Checkerboard patterns (basketweave) are also possible, the smallest of which is known as *seed/moss stitch*: the stitches alternate between knit and purl in every wale and along every row.

Fabrics in which the number of knit and purl stitches are not the same, such as stockinette/stocking stitch, have a tendency to curl; by contrast, those in which knit and purl stitches are arranged symmetrically (such as ribbing, garter stitch or seed/moss stitch) tend to lie flat and drape well. Wales of purl stitches have a tendency to recede, whereas those of knit stitches tend to come forward. Thus, the purl wales in ribbing tend to be invisible, since the neighboring knit wales come forward. Conversely, rows of purl stitches tend to form an embossed ridge relative to a row of knit stitches. This is the basis of shadow knitting, in which the appearance of a knitted fabric changes when viewed from different directions.

Typically, a new stitch is passed through a single unsecured ("active") loop, thus lengthening that wale by one stitch. However, this need not be so; the new loop may be passed through an already secured stitch lower down on the fabric, or even between secured stitches (a dip stitch). Depending on the distance between where the loop is drawn through the fabric and where it is knitted, dip stitches can produce a subtle stippling or long lines across the surface of the fabric, e.g., the lower leaves of a flower. The new loop may also be passed between two stitches in the *present* row, thus clustering the intervening stitches; this approach is often used to produce a smocking effect in the fabric. The new loop may also be passed through *two or more* previous stitches, producing a decrease and merging wales together. The merged stitches need not be from the same row; for example, a tuck can be formed by knitting stitches together from two different rows, producing a raised horizontal welt on the fabric.

Not every stitch in a row need be knitted; some may be left "as is" and knitted on a subsequent row. This is known as slip-stitch knitting. The slipped stitches are naturally longer than the knitted ones. For example, a stitch slipped for one row before knitting would be roughly twice as tall as its knitted counterparts. This can produce interesting visual effects, although the resulting fabric is more rigid because the slipped stitch "pulls" on its neighbours and is less deformable. Mosaic knitting is a form of slip-stitch knitting that knits alternate colored rows and uses slip stitches to form patterns; mosaic-knit fabrics tend to be stiffer than patterned fabrics produced by other methods such as Fair-Isle knitting.

In some cases, a stitch may be deliberately left unsecured by a new stitch and its wale allowed to disassemble. This is known as drop-stitch knitting, and produces a vertical ladder of see-through holes in the fabric, corresponding to where the wale had been.

The stitches on the right are right-plaited, whereas the stitches on the left are left-plaited.

Within limits, an arbitrary number of twists may be added to new stitches, whether they be knit or purl. Here, a single twist is illustrated, with left-plaited and right-plaited stitches on the left and right, respectively.

Right- and Left-plaited Stitches

Both knit and purl stitches may be twisted: usually once if at all, but sometimes twice and (very rarely) thrice. When seen from above, the twist can be clockwise (right yarn over left) or counter-clockwise (left yarn over right); these are denoted as right- and left-plaited stitches, respectively. Hand-knitters generally produce right-plaited stitches by knitting or purling through the back loops, i.e., passing the needle through the initial stitch in an unusual way, but wrapping the yarn as usual. By contrast, the left-plaited stitch is generally formed by hand-knitters by wrapping the yarn in the opposite way, rather than by any change in the needle. Although they are mirror images in form, right- and left-plaited stitches are functionally equivalent. Both types of plaited stitches give a subtle but interesting visual texture, and tend to draw the fabric inwards, making it stiffer. Plaited stitches are a common method for knitting jewelry from fine metal wire.

Illustration of entrelac. The blue and white wales are parallel to each other, but both are perpendicular to the black and gold wales, resembling basket weaving.

Edges and Joins Between Fabrics

The initial and final edges of a knitted fabric are known as the *cast-on* and *bound/cast-off* edges. The side edges are known as the *selvages*; the word derives from "self-edges", meaning that the stitches do not need to be secured by anything else. Many types of selvages have been developed, with different elastic and ornamental properties.

Vertical and horizontal edges can be introduced within a knitted fabric, e.g., for button holes, by binding/casting off and re-casting on again (horizontal) or by knitting the fabrics on either side of a vertical edge separately.

Two knitted fabrics can be joined by embroidery-based grafting methods, most commonly the Kitchener stitch. New wales can be begun from any of the edges of a knitted fabric; this is known as picking up stitches and is the basis for entrelac, in which the wales run perpendicular to one another in a checkerboard pattern.

Illustration of cable knitting. The central braid is formed from 2x2 ribbing in which the background is formed of purl stitches and the cables are each two wales of knit stitches. By changing the order in which the stitches are knit, the wales can be made to cross.

Cables, Increases, and Lace

Ordinarily, stitches are knitted in the same order in every row, and the wales of the fabric run parallel and vertically along the fabric. However, this need not be so, since the order in which stitches are knitted may be permuted so that wales cross over one another, forming a cable pattern. Cables patterns tend to draw the fabric together, making it denser and less elastic; Aran sweaters are a common form of knitted cabling. Arbitrarily complex braid patterns can be done in cable knitting, with the proviso that the wales must move ever upwards; it is generally impossible for a wale to move up and then down the fabric. Knitters have developed methods for giving the illusion of a circular wale, such as appear in Celtic knots, but these are inexact approximations. However, such circular wales are possible using Swiss darning, a form of embroidery, or by knitting a tube separately and attaching it to the knitted fabric.

In lace knitting, the pattern is formed by making small, stable holes in the fabric, generally with yarn overs.

A wale can split into two or more wales using increases, most commonly involving a yarn over. Depending on how the increase is done, there is often a hole in the fabric at the point of the increase. This is used to great effect in lace knitting, which consists of making patterns and pictures using such holes, rather than with the stitches themselves. The large and many holes in lacy knitting makes it extremely elastic; for example, some Shetland "wedding-ring" shawls are so fine that they may be drawn through a wedding ring.

By combining increases and decreases, it is possible to make the direction of a wale slant away from vertical, even in weft knitting. This is the basis for bias knitting, and can be used for visual effect, similar to the direction of a brush-stroke in oil painting.

Ornamentations and Additions

Various point-like ornaments may be added to knitting for their look or to improve the wear of the fabric. Examples include various types of bobbles, sequins and beads. Long loops can also be drawn out and secured, forming a "shaggy" texture to the fabric; this is known as loop knitting. Additional patterns can be made on the surface of the knitted fabric using embroidery; if the embroidery resembles knitting, it is often called Swiss darning. Various closures for the garments, such as frogs and buttons can be added; usually buttonholes are knitted into the garment, rather than cut.

Ornamental pieces may also be knitted separately and then attached using applique. For example, differently colored leaves and petals of a flower could be knit separately and attached to form the final picture. Separately knitted tubes can be applied to a knitted fabric to form complex Celtic knots and other patterns that would be difficult to knit.

Unknitted yarns may be worked into knitted fabrics for warmth, as is done in tufting and "weaving" (also known as "couching").

History and Culture

The word is derived from *knot* and ultimately from the Old English *cnyttan*, to knot.

Nålebinding (Danish: literally "binding with a needle" or "needle-binding") is a fabric creation technique predating both knitting and crochet. That said, one of the earliest known examples of true knitting was cotton socks with stranded knit color patterns found in Egypt from the end of the first millennium AD.

The first commercial knitting guilds appear in Western Europe in the early fifteenth century (Tournai in 1429, Barcelona in 1496). The Guild of Saint Fiacre was founded in Paris in 1527 but the archives mention an organization (not necessarily a guild) of knitters from 1268.

With the invention of the knitting machine, knitting "by hand" became a craft used by country people with easy access to fiber. Similar to quilting, spinning, and needlepoint, hand knitting became a leisure activity for the wealthy.

Properties of Fabrics

The topology of a knitted fabric is relatively complex. Unlike woven fabrics, where strands usually run straight horizontally and vertically, yarn that has been knitted follows a looped path along its row, as with the red strand in the diagram at left, in which the loops of one row have all been pulled through the loops of the row below it.

Close-up of back of stockinette stitch, also same appearance as reverse stockinette stitch

Because there is no single straight line of yarn anywhere in the pattern, a knitted piece of fabric can stretch in all directions. This elasticity is all but unavailable in woven fabrics which only stretch along the bias. Many modern stretchy garments, even as they rely on elastic synthetic materials for some stretch, also achieve at least some of their stretch through knitted patterns.

The basic knitted fabric (as in the diagram, and usually called a *stocking* or *stockinette* pattern) has a definite "right side" and "wrong side". On the right side, the visible portions of the loops are the verticals connecting two rows which are arranged in a grid of *V* shapes. On the wrong side, the ends of the loops are visible, both the tops and bottoms, creating a much more bumpy texture sometimes called *reverse stockinette*. (Despite being the "wrong side," reverse stockinette is frequently used as a pattern in its own right.) Because the yarn holding rows together is all on the front, and the yarn holding side-by-side stitches together is all on the back, stockinette fabric has a strong tendency to curl toward the front on the top and bottom, and toward the back on the left and right side.

Stitches can be worked from either side, and various patterns are created by mixing regular knit stitches with the "wrong side" stitches, known as purl stitches, either in columns (ribbing), rows (garter, welting), or more complex patterns. Each fabric has different properties: a garter stitch has much more vertical stretch, while ribbing stretches much more horizontally. Because of their front-back symmetry, these two fabrics have little curl, making them popular as edging, even when their stretch properties are not desired.

Different combinations of knit and purl stitches, along with more advanced techniques, generate fabrics of considerably variable consistency, from gauzy to very dense, from highly stretchy to relatively stiff, from flat to tightly curled, and so on.

Texture

The most common texture for a knitted garment is that generated by the flat stockinette stitch—as seen, though very small, in machine-made stockings and T-shirts—which is worked in the round as nothing but knit stitches, and worked flat as alternating rows of knit and purl. Other simple textures can be made with nothing but knit and purl stitches, including garter stitch, ribbing, and moss and seed stitches. Adding a "slip stitch" (where a loop is passed from one needle to the other) allows for a wide range of textures, including heel and linen stitches as well as a number of more complicated patterns.

Close-up of ribbing

Some more advanced knitting techniques create a surprising variety of complex textures. Combining certain increases, which can create small eyelet holes in the resulting fabric, with assorted decreases is key to creating knitted lace, a very open fabric resembling lace. Open vertical stripes can be created using the drop-stitch knitting technique. Changing the order of stitches from one row to the next, usually with the help of a cable needle or stitch holder, is key to cable knitting, producing an endless variety of cables, honeycombs, ropes, and Aran sweater patterning. Entrelac forms a rich checkerboard texture by knitting small squares, picking up their side edges, and knitting more squares to continue the piece.

Fair Isle knitting uses two or more colored yarns to create patterns and forms a thicker and less flexible fabric.

The appearance of a garment is also affected by the *weight* of the yarn, which describes the thickness of the spun fibre. The thicker the yarn, the more visible and apparent stitches will be; the thinner the yarn, the finer the texture.

Color

Plenty of finished knitting projects never use more than a single color of yarn, but there are many ways to work in multiple colors. Some yarns are dyed to be either *variegated* (changing color every few stitches in a random fashion) or *self-striping* (changing every few rows). More complicated techniques permit large fields of color (intarsia, for example), busy small-scale patterns of color (such as Fair Isle), or both (double knitting and slip-stitch color, for example).

Yarn with multiple shades of the same hue are called *ombre*, while a yarn with multiple hues may be known as a given *colorway* — a green, red and yellow yarn might be dubbed the "Parrot Colorway" by its manufacturer, for example. *Heathered* yarns contain small amounts of fibre of different colours, while *tweed* yarns may have greater amounts of different colored fibres.

Hand Knitting Process

A woman in the process of hand knitting (1904)

There are many hundreds of different knitting stitches used by hand knitters. A piece of hand knitting begins with the process of *casting on*, which involves the initial creation of the stitches on the needle. Different methods of casting on are used for different effects: one may be stretchy enough for lace, while another provides a decorative edging. *Provisional* cast-ons are used when the knitting will continue in both directions from the cast-on. There are various methods employed to cast on, such as the "thumb method" (also known as "slingshot" or "long-tail" cast-ons), where the stitches are created by a series of loops that will, when knitted, give a very loose edge ideal for "picking up stitches" and knitting a border; the "double needle method" (also known as "knit-on" or "cable cast-on"), whereby each loop placed on the needle is then "knitted on," which produces a firmer edge ideal on its own as a border; and many more. The number of active stitches remains the same as when cast on unless stitches are added (an increase) or removed (a decrease).

Most Western-style hand knitters follow either the English style (in which the yarn is held in the right hand) or the Continental style (in which the yarn is held in the left hand).

There are also different ways to insert the needle into the stitch. Knitting through the front of a stitch is called Western knitting. Going through the back of a stitch is called Eastern knitting. A third method, called combination knitting, goes through the front of a knit stitch and the back of a purl stitch.

Once the hand knitted piece is finished, the remaining live stitches are "cast off". Casting (or "binding") off loops the stitches across each other so they can be removed from the needle without unravelling the item. Although the mechanics are different from casting on, there is a similar variety of methods.

In hand knitting certain articles of clothing, especially larger ones like sweaters, the final knitted garment will be made of several knitted pieces, with individual sections of the garment hand knitted separately and then sewn together. Seamless knitting, where a whole garment is hand knit as a single piece, is also possible. Elizabeth Zimmermann is probably the best-known proponent of seamless or circular hand knitting techniques. Smaller items, such as socks and hats, are usually knit in one piece on double-pointed needles or circular needles. Hats in particular can be started "top down" on double pointed needles with the increases added until the preferred size is achieved, switching to an appropriate circular needle when enough stitches have been added. Care must be taken to bind off at a tension that will allow the "give" needed to comfortably fit on the head.

Mega Knitting

Mega knitting is a term recently coined and relates to the use of knitting needles greater than or equal to half an inch in diameter.

Mega knitting uses the same stitches and techniques as conventional knitting, except that hooks are carved into the ends of the needles. The hooked needles greatly enhance control of the work, catching the stitches and preventing them from slipping off.

It was the development of the knitting machine that introduced hooked needles and enabled faultless, automated knitting. Most knitters probably aren't even aware of the many processes that their fingers perform in the making of a single stitch. However, large gauge needles emphasise those actions and knitting becomes increasingly more awkward when the needle diameter is greater than the width of the knitters finger. On a one-inch diameter (size 50) needle for instance, the shaft begins to taper one and three quarter inches from the tip. This means that the stitches are spread much further apart on mega knitting needles, making them more difficult to control. The hook catches the loop of yarn as each stitch is knitted, meaning that wrists and fingers don't have to work so hard and there is less chance of stitches slipping off the needle. The position of the hook is most important. Turn the left (non-working) hook to face away at all times; turn the right (working) hook toward you up whilst knitting (plain stitch) and away whilst purling.

Mega knitting produces a chunky, bulky fabric or an open lacy weave, depending on the weight and type of yarn used.

Materials

Yarn

A hank of wool yarn (center) is uncoiled into its basic loop. A tie is visible at the left; after untying, the hank may be wound into a ball or balls suitable for knitting. Knitting from a normal hank directly is likely to tangle the yarn, producing snarls.

Yarn for hand-knitting is usually sold as balls or skeins (hanks), although it may also be wound on spools or cones. Skeins and balls are generally sold with a *yarn-band*, a label that describes the yarn's weight, length, dye lot, fiber content, washing instructions, suggested needle size, likely gauge/tension, etc. It is common practice to save the yarn band for future reference, especially if additional skeins must be purchased. Knitters generally ensure that the yarn for a project comes from a single dye lot. The dye lot specifies a group of skeins that were dyed together and thus have precisely the same color; skeins from different dye-lots, even if very similar in color, are usually slightly different and may produce a visible horizontal stripe when knitted together. If a knitter buys insufficient yarn of a single dye lot to complete a project, additional skeins of the same dye lot can sometimes be obtained from other yarn stores or online. Otherwise, knitters can alternate skeins every few rows to help the dye lots blend together easier.

Transformation of a hank of lavender silk yarn (top) into a ball in which the knitting yarn emerges from the center (bottom). The latter is better for knitting, since the yarn is much less likely to tangle.

The thickness or weight of the yarn is a significant factor in determining the gauge/tension, i.e., how many stitches and rows are required to cover a given area for a given stitch pattern. Thicker yarns generally require thicker knitting needles, whereas thinner yarns may be knit with thick or thin needles. Hence, thicker yarns generally require fewer stitches, and therefore less time, to knit up a given garment. Patterns and motifs are coarser with thicker yarns; thicker yarns produce bold visual effects, whereas thinner yarns are best for refined patterns. Yarns are grouped by thickness into six categories: superfine, fine, light, medium, bulky and superbulky; quantitatively, thickness is measured by the number of wraps per inch (WPI). In the British Commonwealth (outside North America) yarns are measured as 1ply, 2ply, 3ply, 4ply, 5ply, 8ply (or double knit),10ply and 12ply (triple knit). The related *weight per unit length* is usually measured in tex or denier.

Before knitting, the knitter will typically transform a hank/skein into a ball where the yarn emerges from the center of the ball; this making the knitting easier by preventing the yarn from becoming easily tangled. This transformation may be done by hand, or with a device known as a ballwinder. When knitting, some knitters enclose their balls in jars to keep them clean and untangled with other yarns; the free yarn passes through a small hole in the jar-lid.

A yarn's usefulness for a knitting project is judged by several factors, such as its *loft* (its ability to trap air), its *resilience* (elasticity under tension), its washability and colorfastness, its *hand* (its feel, particularly softness vs. scratchiness), its durability against abrasion, its resistance to pilling, its *hairiness* (fuzziness), its tendency to twist or untwist, its overall weight and drape, its blocking and felting qualities, its comfort (breathability, moisture absorption, wicking properties) and of course its look, which includes its color, sheen, smoothness and ornamental features. Other factors include allergenicity; speed of drying; resistance to chemicals, moths, and mildew; melting point and flammability; retention of static electricity; and the propensity to become stained and to accept dyes. Different factors may be more significant than others for different knitting projects, so there is no one "best" yarn. The resilience and propensity to (un)twist are general properties that affect the ease of hand-knitting. More resilient yarns are more forgiving of irregularities in tension; highly twisted yarns are sometimes difficult to knit, whereas untwisting yarns can lead to split stitches, in which not all the yarn is knitted into a stitch. A key factor in knitting is *stitch definition*, corresponding to how well complicated stitch patterns can be seen when made from a given yarn. Smooth, highly spun yarns are best for showing off stitch patterns; at the other extreme, very fuzzy yarns or eyelash yarns have poor stitch definition, and any complicated stitch pattern would be invisible.

Although knitting may be done with ribbons, metal wire or more exotic filaments, most yarns are made by spinning fibers. In spinning, the fibers are twisted so that the yarn resists breaking under tension; the twisting may be done in either direction, resulting in a Z-twist or S-twist yarn. If the fibers are first aligned by combing them, the yarn is smoother and called a *worsted*; by contrast, if the fibers are carded but not combed, the yarn is fuzzier and called *woolen-spun*. The fibers making up a yarn may be continuous *filament* fibers such as silk and many synthetics, or they may be *staples* (fibers of an average length, typically a few inches); naturally filament fibers are sometimes cut up into staples before spinning. The strength of the spun yarn against breaking is determined by the amount of twist, the length of the fibers and the thickness of the yarn. In general, yarns become stronger with more twist (also called *worst*), longer fibers and thicker yarns (more fibers); for example, thinner yarns require more twist than do thicker yarns to resist breaking under tension. The thickness of the yarn may vary along its length; a *slub* is a much thicker section in which a mass of fibers is incorporated into the yarn.

The spun fibers are generally divided into animal fibers, plant and synthetic fibers. These fiber types are chemically different, corresponding to proteins, carbohydrates and synthetic polymers, respectively. Animal fibers include silk, but generally are long hairs of animals such as sheep (wool), goat (angora, or cashmere goat), rabbit (angora), llama, alpaca, dog, cat, camel, yak, and muskox (qiviut). Plants used for fibers include cotton, flax (for linen), bamboo, ramie, hemp, jute, nettle, raffia, yucca, coconut husk, banana fiber, soy and corn. Rayon and acetate fibers are also produced from cellulose mainly derived from trees. Common synthetic fibers include acrylics, polyesters such as dacron and ingeo, nylon and other polyamides, and olefins such as polypropylene. Of these types, wool is generally favored for knitting, chiefly owing to its superior elasticity, warmth and (sometimes) felting; however, wool is generally less convenient to clean and some people are allergic to it. It is also common to blend different fibers in the yarn, e.g., 85% alpaca and 15% silk. Even within a type of fiber, there can be great variety in the length and thickness of the fibers; for example, Merino wool and Egyptian cotton are favored because they produce exceptionally long, thin (fine) fibers for their type.

A single spun yarn may be knitted as is, or braided or plied with another. In plying, two or more yarns are spun together, almost always in the opposite sense from which they were spun individually; for example, two Z-twist yarns are usually plied with an S-twist. The opposing twist relieves some of the yarns' tendency to curl up and produces a thicker, *balanced* yarn. Plied yarns may themselves be plied together, producing *cabled yarns* or *multi-stranded yarns*. Sometimes, the yarns being plied are fed at different rates, so that one yarn loops around the other, as in bouclé. The single yarns may be dyed separately before plying, or afterwards to give the yarn a uniform look.

The dyeing of yarns is a complex art that has a long history. However, yarns need not be dyed. They may be dyed just one color, or a great variety of colors. Dyeing may be done industrially, by hand or even hand-painted onto the yarn. A great variety of synthetic dyes have been developed since the synthesis of indigo dye in the mid-19th century; however, natural dyes are also possible, although they are generally less brilliant. The color-scheme of a yarn is sometimes called its colorway. Variegated yarns can produce interesting visual effects, such as diagonal stripes; conversely, a variegated yarn may frustrate an otherwise good knitting pattern by producing distasteful color combination.

Glass/Wax

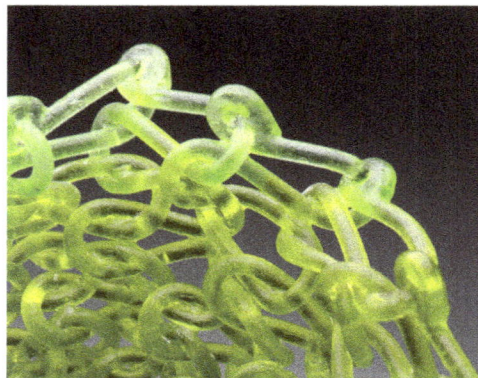

Close-up of 'Jitterbug' - Knitted Glass

Knitted Glass combines knitting, lost-wax casting, mold-making, and kiln-casting. The process involves

1. *knitting* with wax strands,

2. surrounding the knitted wax piece with a heat-tolerant refractory material,

3. removing the wax by melting it out, thus creating a mold;

4. placing the mold in a kiln where lead crystal glass melts into the mold;

5. after the mold cools, the mold material is removed to reveal the knitted glass piece.

Tools

The process of knitting has three basic tasks:

1. the active (unsecured) stitches must be held so they don't drop

2. these stitches must be released sometime after they are secured

3. new bights of yarn must be passed through the fabric, usually through active stitches, thus securing them.

In very simple cases, knitting can be done without tools, using only the fingers to do these tasks; however, knitting is usually carried out using tools such as knitting needles, knitting machines or rigid frames. Depending on their size and shape, the rigid frames are called stocking frames, knitting boards, knitting rings (also called knitting looms) or knitting spools (also known as knitting knobbies, knitting nancies, or corkers). There is also a technique called knooking of knitting with a crochet hook that has a cord attached to the end, to hold the stitches while they're being worked. Other tools are used to prepare yarn for knitting, to measure and design knitted garments, or to make knitting easier or more comfortable.

Needles

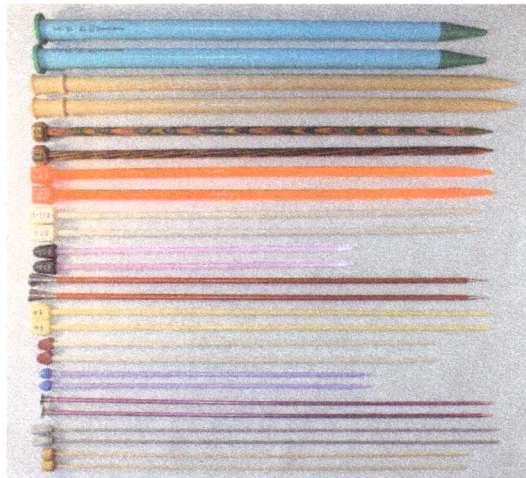

Knitting needles in a variety of sizes and materials. Different materials have varying amounts of friction, and are suitable for different yarn types.

There are three basic types of knitting needles (also called "knitting pins"). The first and most common type consists of two slender, straight sticks tapered to a point at one end, and with a knob at the other end to prevent stitches from slipping off. Such needles are usually 10–16 inches

(250–410 mm) long but, due to the compressibility of knitted fabrics, may be used to knit pieces significantly wider. The most important property of needles is their diameter, which ranges from below 2 to 25 mm (roughly 1 inch). The diameter affects the size of stitches, which affects the gauge/tension of the knitting and the elasticity of the fabric. Thus, a simple way to change gauge/tension is to use different needles, which is the basis of uneven knitting. Although the diameter of the knitting needle is often measured in millimeters, there are several measurement systems, particularly those specific to the United States, the United Kingdom and Japan; a conversion table is given at knitting needle. Such knitting needles may be made out of any materials, but the most common materials are metals, wood, bamboo, and plastic. Different materials have different frictions and grip the yarn differently; slick needles such as metallic needles are useful for swift knitting, whereas rougher needles such as bamboo offer more friction and are therefore less prone to dropping stitches. The knitting of new stitches occurs only at the tapered ends. Needles with lighted tips have been sold to allow knitters to knit in the dark.

Double-pointed knitting needles in various materials and sizes. They come in sets of four, five or six.

The second type of knitting needles are straight, double-pointed knitting needles (also called "DPNs"). Double-pointed needles are tapered at both ends, which allows them to be knit from either end. DPNs are typically used for circular knitting, especially smaller tube-shaped pieces such as sleeves, collars, and socks; usually one needle is active while the others hold the remaining active stitches. DPNs are somewhat shorter (typically 7 inches) and are usually sold in sets of four or five.

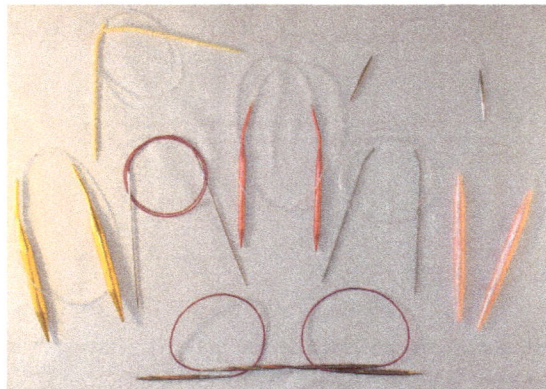

Circular knitting needles in different lengths, materials and sizes, including plastic, aluminum, steel and nickel-plated brass

Cable needles are a special case of DPNs, although they are usually not straight, but dimpled in the middle. Often, they have the form of a hook. When cabling a knitted piece, a hook is easier to grab and hold the yarn. Cable needles are typically very short (a few inches), and are used to hold stitches temporarily while others are being knitted. Cable patterns are made by permuting the order of stitches; although one or two stitches may be held by hand or knit out of order, cables of three or more generally require a cable needle.

The third needle type consists of circular needles, which are long, flexible double-pointed needles. The two tapered ends (typically 5 inches (130 mm) long) are rigid and straight, allowing for easy knitting; however, the two ends are connected by a flexible strand (usually nylon) that allows the two ends to be brought together. Circular needles are typically 24-60 inches long, and are usually used singly or in pairs; again, the width of the knitted piece may be significantly longer than the length of the circular needle.

A developing trend in the knitting world is interchangeable needles. These kits consist of pairs of needles with usually nylon cables or cords. The cables/cords are screwed into the needles, allowing the knitter to have both flexible straight needles or circular needles. This also allows the knitter to change the diameter and length of the needles as needed.

The ability to work from either end of one needle is convenient in several types of knitting, such as slip-stitch versions of double knitting. Circular needles may be used for flat or circular knitting.

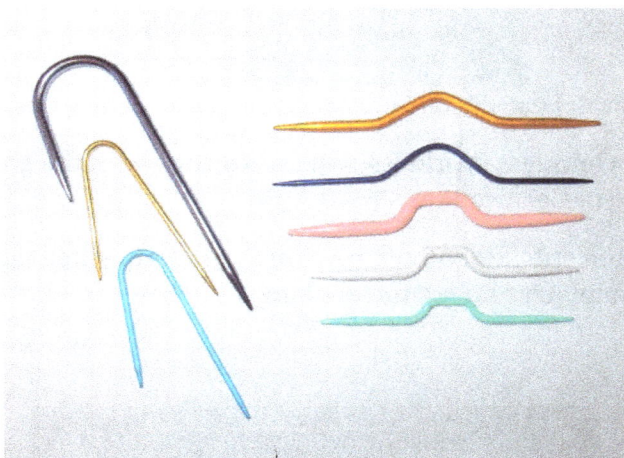

Cable needles

Cable needles are a specific design, and are used to create the twisting motif of a knitted cable. They are made in different sizes, which produces cables of different widths. When in use, the cable needle is used at the same time as two regular needles. It functions by holding together the stitches creating the cable as the other needles create the rest of the stitches for the knitted piece. At specific points indicated by the knitting pattern, the cable needle is moved, the stitches on it are worked by the other needles, then the cable needle is turned around to a different position to create the cable twist.

Mega Needles

Mega knitting needles are generally considered to be any knitting needles larger than size 17 (half inch diameter). Mega needles may or may not have hooks carved in the ends. Hooks on large diameter needles help enormously to control the stitches whilst knitting.

Largest Circular Knitting Needles

The largest aluminum circular knitting needles on record are size US 150 and are nearly 7 feet tall. They are owned by Paradise Fibers and are currently on display in the Paradise Fibers retail showroom.

Record

Julia Hopson with world-record 3.5 meter long knitting needles

The current holder of the Guinness World Record for Knitting with the Largest Knitting Needles is Julia Hopson of Penzance in Cornwall.

Julia knitted a square of ten stitches and ten rows in stockinette stitch using knitting needles that were 6.5 centimeters in diameter and 3.5 meters long.

Ancillary Tools

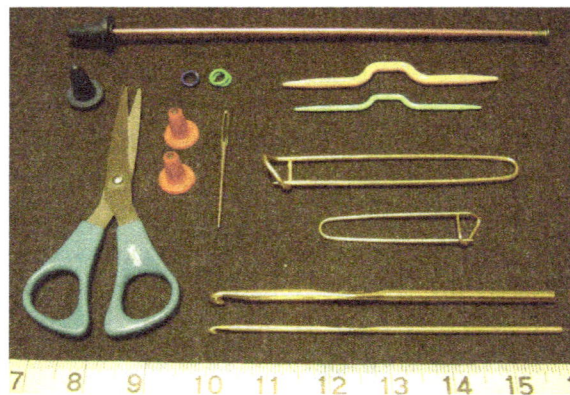

Some ancillary tools used by hand-knitters. Starting from the bottom right are two crochet hooks, two stitch holders (like big blunt safety pins), and two cable needles in pink and green. On the left are a pair of scissors, a yarn needle, green and blue stitch markers, and two orange point protectors. At the top left are two blue point protectors, one on a red needle.

Various tools have been developed to make hand-knitting easier. Tools for measuring needle diameter and yarn properties have been discussed above, as well as the yarn swift, ballwinder and "yarntainers". Crochet hooks and a darning needle are often useful in binding/casting off or in joining two knitted pieces edge-to-edge. The darning needle is used in duplicate stitch (also known as Swiss darning). The crochet hook is also essential for repairing dropped stitches and some specialty stitches such as tufting. Other tools such as knitting spools or pom-pom makers are used to prepare specific ornaments. For large or complex knitting patterns, it is sometimes difficult to keep track of which stitch should be knit in a particular way; therefore, several tools have been developed to identify the number of a particular row or stitch, including circular stitch markers, hanging markers, extra yarn and row counters. A second potential difficulty is that the knitted piece will slide off the tapered end of the needles when unattended; this is prevented by "point protectors" that cap the tapered ends. Another problem is that too much knitting may lead to hand and wrist troubles; for this, special stress-relieving gloves are available. In traditional Shetland knitting a special belt is often used to support the end of one needle allowing the knitting greater speed. Finally, there are sundry bags and containers for holding knitting, yarns and needles.

Commercial Applications

Industrially, metal wire is also knitted into a metal fabric for a wide range of uses including the filter material in cafetieres, catalytic converters for cars and many other uses. These fabrics are usually manufactured on circular knitting machines that would be recognised by conventional knitters as sock machines.

Many fashion designers make heavy use of knitted fabric in their fashion collections. Gordana Gelhausen, who appeared in season six of the television show *Project Runway*, is primarily a knit designer. Other designers and labels that make heavy use of knitting include Michael Kors, Fendi, and Marc Jacobs.

For individual hobbyists, websites such as Etsy, Big Cartel and Ravelry have made it easy to sell knitting patterns on a small scale, in a way similar to eBay.

Graffiti

In the last decade, a practice called knitting graffiti, guerilla knitting, or yarn bombing— the use of knitted or crocheted cloth to modify and beautify one's (usually outdoor) surroundings—emerged in the U.S. and spread worldwide. Magda Sayeg is credited with starting the movement in the US and Knit the City are a prominent group of graffiti knitters in the United Kingdom. Yarn bombers sometimes target existing pieces of graffiti for beautification. For instance, Dave Cole is a contemporary sculpture artist who practiced knitting as graffiti for a large-scale public art installation in Melbourne, Australia for the Big West Arts Festival in 2009. The work was vandalized the night of its completion. A new movie, shot by a Tasmanian filmmaker on a set made almost entirely out of yarn, was partially inspired by "knitted graffiti".

Yarn Crawl

Many major metropolitan cities across the US and Europe host annual Yarn Crawls. The event is typically a multi-day event that caters to all knitters, crochet and yarn enthusiasts that supports the local crafting community. Over the multi-day period, multiple local yarn and knit shops participate in the yarn crawl and offer up store discounts, give away free exclusive patterns, provide classes, trunk shows and conduct raffles for prizes. Participants of the crawl receive a passport and get their passport stamped at each store they visit along the crawl. Traditionally those that get their passports fully stamped are eligible to win a larger gift basket filled with yarn, knitting and crochet goodies. Some local crawls also provide a Knit-Along (KAL) or Crochet-Along (CAL) where attendees follow a specific pattern prior to the crawl and then proudly wear it during the crawl for others.

Charity

Hand knitting garments for free distribution to others has become common practice among hand knitting groups. Girls and women hand knitted socks, sweaters, scarves, mittens, gloves, and hats for soldiers in Crimea, the American Civil War, and the Boer Wars; this practice continued in World War I, World War II and the Korean War, and continues for soldiers in Iraq and Afghanistan. The Australian charity *Wrap with Love* continues to provide blankets hand knitted by volunteers to people most in need around the world who have been affected by war.

In the historical projects, yarn companies provided knitting patterns approved by the various branches of the armed services; often they were distributed by local chapters of the American Red Cross. Modern projects usually entail the hand knitting of hats or helmet liners; the liners provided for soldiers must be of 100% worsted weight wool and be crafted using specific colors.

Some charities teach women to knit as a means of clothing their families or supporting themselves.

Clothing and afghans are frequently made for children, the elderly, and the economically disadvantaged in various countries. Pine Ridge Indian Reservation accepts donations for the Lakota people in the United States. Prayer shawls, or shawls in which the crafter meditates or says prayers of their faith while hand knitting with the intent on comforting the recipient, are donated to those experiencing loss or stress. Many knitters today hand knit and donate "chemo caps," soft caps for cancer patients who lose their hair during chemotherapy. Yarn companies offer free knitting patterns for these caps.

Penguin sweaters were hand knitted by volunteers for the rehabilitation of penguins contaminated by exposure to oil slicks. The project is now complete.

Chicken sweaters were also hand knitted to aid battery hens that had lost their feathers. The organization is not currently accepting donations, but maintains a list of volunteers.

Originally started after the 2004 Indonesian tsunami, Knitters Without Borders is a charity challenge issued by knitting personality Stephanie Pearl-McPhee that encourages hand knitters to donate to Médecins Sans Frontières (Doctors Without Borders). Instead of hand knitting for charity, knitters are encouraged to donate a week's worth of disposable income, including money that otherwise might have been spent on yarn. Knitted items are occasional offered as prizes to donors. As of September 2011, Knitters Without Borders donors have contributed CAD$1,062,217.

Security blankets can also be made through the Project Linus organization which helps needy children.

There are organizations that help reach other countries in need such as afghans for Afghans. This outreach is described as, "afghans for Afghans is a humanitarian and educational people-to-people project that sends hand-knit and crocheted blankets and sweaters, vests, hats, mittens, and socks to the beleaguered people of Afghanistan."

Health Benefits

Studies have shown that hand knitting, along with other forms of needlework, provide several significant health benefits. These studies have found the rhythmic and repetitive action of hand knitting can "help prevent and manage stress, pain and depression, which in turn strengthens the body's immune system", as well as create a relaxation response in the body which can decrease blood pressure, heart rate, help prevent illness, and have a calming effect. Pain specialists have also found that the brain chemistry is changed when one hand knits, resulting in an increase in "feel good" hormones (i.e. serotonin and dopamine), and a decrease in stress hormones.

Hand knitting, along with other leisure activities has been linked to reducing the risk of developing Alzheimer's disease and dementia. Much like physical activity strengthens the body, mental exercise makes the human brain more resilient.

A repository of research into the effect on health of hand knitting can be found at Stitchlinks, an organisation founded in Bath, England.

Knitting also helps in the area of social interaction; knitting provides people with opportunities to socialize with others. Some ways to increase social interaction with knitting is inviting friends over to knit and chat with each other. Even if they've never knitted before this can be a fun way to interact with your friends.

Another interesting way that knitting can positively impact your life is improving the dexterity in your hands and figures. This keeps your fingers limber and can be especially helpful for those with arthritis. Knitting can reduce the pain of arthritis if people make it a daily habit.

Lace Knitting

Lace knitting is a style of knitting characterized by stable "holes" in the fabric arranged with consideration of aesthetic value. Lace is sometimes considered the pinnacle of knitting, because of its complexity and because woven fabrics cannot easily be made to have holes. Famous examples include the wedding ring shawl of Shetland knitting, a shawl so fine that it could be drawn through a wedding ring. Shetland knitted lace became extremely popular in Victorian England when Queen Victoria became a Shetland lace enthusiast. From there, knitting patterns for the shawls were printed in English women's magazines where they were copied in Iceland with single ply wool.

Some consider that "true" knitted lace has pattern stitches on both the right and wrong sides, and that knitting with pattern stitches on only one side of the fabric, so that holes are separated by at least two threads, is technically not lace, but "lacy knitting", although this has no historical basis.

Eyelet patterns are those in which the holes make up only a small fraction of the fabric and are isolated into clusters (e.g., little rosettes of one hole surrounded by others in a hexagon). At the other extreme, some knitted lace is almost all holes, e.g., faggoting.

Rectangular lace shawl on the needles. White threads ("lifelines") are strung through the pattern every twenty rows and will be removed upon completion.

Knitted lace with no bound-off edges is extremely elastic, deforming easily to fit whatever it is draped on. As a consequence, knitted lace garments must be blocked or "dressed" before use, and tend to stretch over time.

Lace can be used for any kind of garment, but is commonly associated with scarves and shawls, or with household items such as curtains, table runners or trim for curtains and towels. Lace items from different regional knitting traditions are often distinguished by their patterns, shape and method, such as Faroese lace shawls which are knit bottom up with center back gusset shaping unlike a more common neck down, triangular shawl.

Technique

A hole can be introduced into a knitted fabric by pairing a yarn over stitch with a nearby (usually adjacent) decrease. If the decrease *precedes* the yarn over, it typically slants *right* as seen from the right side (e.g., k2tog, *not* k2tog tbl; see knitting abbreviations). If the decrease *follows* the yarn over, it typically slants *left* as seen from the right side (e.g., k2tog tbl or ssk, *not* k2tog). These slants pull the fabric away from the yarn over, opening up the hole.

Lace scarf during blocking

Pairing a yarn over with a decrease keeps the stitch count constant. Many beautiful patterns separate the yarn over and decrease stitches, e.g., k2tog, k5, yo. Separating the yarn over from its decrease "tilts" all the intervening stitches towards the decrease. The tilt may form part of the design, e.g., mimicking the veins in a leaf.

There are few constraints on positioning the holes, so practically any picture or pattern can be outlined with holes; common motifs include leaves, rosettes, ferns and flowers. To design a simple lace motif, a knitter can draw its lines on a piece of knitting graph paper; right-slanting lines should be produced with "k2tog, yo" stitch-pairs (as seen on the right side) whereas left-slanting lines should be produced with "yo, k2tog tbl" (or, equivalently, "yo, ssk" or "yo, skp") stitch pairs (again, as seen on the right side). More sophisticated patterns will change the grain of the fabric to help the design, by separating the yarn overs and decreases.

It is common for lace knitters to insert a "lifeline", a strand of contrasting yarn threaded through stitches on the needle, at the end of every pattern repeat or after a certain number of rows. This allows the knitter to rip out a controlled number of rows if a mistake is discovered.

Simple Examples

A horizontal row of holes can be produced by the pattern: *k3, k2tog, yo, k3*.

A pair of vertical columns can be produced by stacking the pattern: (k, k2tog, yo, k, yo, k2tog tbl, k) on the right side. Here the flanking decreases slant outwards away from the central stitch. For a thicker central column, one can move the decreases so that they slant inwards: (k, yo, dec 2 symmetrically, yo, k). For making the same pattern on the wrong side, the converse stitch patterns are: (p, p2tog, yo, p, yo, p2tog tbl, p) and (p, yo, dec 2 symmetrically, yo, p), respectively.

A diagonal row of holes can be made by shifting the (yo, dec) every row or every other row, e.g.,

- Row 1: k, k2tog, yo, k5

- Row 3: k3, k2tog, yo, k3

- Row 5: k5, k2tog, yo, k1

History and Comparison to Other Laces

Lace knitting is generally not as fine as other forms of lace, such as needle lace or bobbin lace. However, it is better suited for garments, being softer and much faster to produce.

Plaited Stitch (Knitting)

In knitting, a plaited stitch is a single knitted stitch that is twisted clockwise (right over left) or counterclockwise (left over right), usually by one half-turn (180°) but sometimes by a full turn (360°) or more.

Methods

Plaited stitches can be produced in several ways. Knitting into the back loop produces a clockwise plaited stitch in the lower stitch being knitted (i.e., the loop that was on the left-hand needle.) The clockwise-plaited stitch is also called a left crossed stitch, since the left strand (i.e., the outgoing strand) of the loop crosses over the right incoming strand. Left-crossed stitches are sometimes called twisted stitches, although the latter term might be confused with similar terms from cable knitting. Conversely, a counterclockwise plaited stitch can be produced if the yarn is wrapped around the needle in the opposite direction as normal while knitting a stitch. Such a stitch is also called a "right crossed stitch", since the right incoming strand crosses over the left outgoing strand. Here, the plait appears in the upper stitch being knitted, i.e., in the new loop being formed. In the "brute-force" approach, the knitter can produce any sort of plaiting by removing the stitch to be knitted from the left-hand needle, twisting it as desired, then returning it to the left-hand needle and knitting it.

Applications in Knitting

Both clockwise and counterclockwise plaited stitches are often repeated in wales, i.e., in columns of knitting, where they make attractive, subtly different ribbings. Fabrics made with plaited stitches are stiffer than normal and "draw in" sideways, i.e., have a smaller widthwise gauge.

Extra-long, full-turn clockwise plaited stitches can be made by knitting through the back loop and wrapping the yarn twice; this is an attractive stitch when repeated in a row, creating openness and a change in scale that enlivens even simple stockinette or garter stitch.

Plaited stitches are also useful in increases and decreases, both for drawing the fabric together and for covering potential "holes" in the fabric.

As a Method for Correcting Errors

As an aside, knitting through the back loop is a useful technique for untwisting stitches on the left-hand needle that "hang backwards". Such stitches are often produced when a knitted fabric is partially pulled out and some stitches are accidentally put back onto the needle with a backwards twist.

Warp Knitting

Warp knitting is a family of knitting methods in which the yarn zigzags along the length of the fabric, i.e., following adjacent columns ("wales") of knitting, rather than a single row ("course"). For comparison, knitting across the width of the fabric is called weft knitting.

Since warp knitting requires that the number of separate strands of yarn ("ends") equals the number of stitches in a row, warp knitting is almost always done by machine, not by hand.

Basic pattern of warp knitting. Parallel yarns zigzag lengthwise along the fabric, each loop securing a loop of an adjacent strand from the previous row.

Types

Warp knitting comprises several types of knitted fabrics. All warp-knit fabrics are resistant to runs and relatively easy to sew. Raschel lace—the most common type of machine made lace—is a warp knit fabric but using many more guide-bars (12+) than the usual machines which mostly have three or four bars.

Tricot

Tricot is very common in lingerie. The right side of the fabric has fine lengthwise ribs while the reverse has crosswise ribs. The properties of these fabrics include having a soft and 'drapey' texture with some lengthwise stretch and almost no crosswise stretch.

Milanese Knit

Milanese is stronger, more stable, smoother and more expensive than tricot and, hence, is used in better lingerie. These knit fabrics are made from two sets of yarn knitted diagonally, which results in the face fabric having a fine vertical rib and the reverse having a diagonal structure, and results in these fabrics being lightweight, smooth, and run-resistant. Milanese is now virtually obsolete.

Raschel Knit

Raschel knits do not stretch significantly and are often bulky; consequently, they are often used as an unlined material for coats, jackets, straight skirts and dresses. These fabrics can be made out of conventional or novelty yarns which allows for interesting textures and designs to be created. The qualities of these fabrics range from "dense and compact to open and lofty [and] can be either stable or stretchy, and single-faced or reversible. The largest outlet for the Raschel Warp

Knitting Machine is for lace fabric and trimmings. Raschel knitting is also used in outdoors and military fabrics for products such as backpacks. It is used to provide a ventilated mesh next to the user's body (covering padding) or mesh pockets and pouches to facilitate visibility of the contents (MIL-C-8061).

In 1855 Redgate combined the principles of a circular loom with those of warp knit. A German firm used this machine to produce "Raschel" shawls, named after the French actress Élisabeth Félice *Rachel*. In 1859 Wilhelm Barfuss improved the machine to create the Raschel machines. The Jacquard apparatus was adapted to it in the 1870s. The Raschel machine could work at higher speeds than the Leavers machine and proved the most adaptable to the new synthetic fibres, such as nylon and polyester, in the 1950s. Most contemporary machine-made lace is made on Raschel machines.

Stitch-bonding

- Stitch-Bonding is a special form of warp knitting and is commonly used for the production of composite materials and technical textiles. As a method of production, stitch-bonding is efficient, and is one of the most modern ways to create reinforced textiles and composite materials for industrial use. The advantages of the stitch-bonding process includes its high productivity rate and the scope it offers for functional design of textiles, such as fiber-reinforced plastics. Stitch-bonding involves layers of threads and fabric being joined together with a knitting thread, which creates a layered structure called a multi-ply. This is created through a warp-knitting thread system, which is fixed on the reverse side of the fabric with a sinker loop, and a weft thread layer. A needle with the warp thread passes through the material, which requires the warp and knitting threads to be moving both parallel and perpendicular to the vertical/warp direction of the stitch-bonding machine. Stitch-bonded fabrics are currently being used in such fields as wind energy generation and aviation. Research is currently being conducted into the usage and benefits of stitch-bonded fabrics as a way to reinforce concrete. Fabrics produced with this process offer the potential of using "sensitive fiber materials such as glass and carbon with only little damage, non-crimp fiber orientation and variable distance between threads".

- Extended Stitch-Bonding process (or the extended warp-knitting process): the compound needle that pierces the piles is shifted laterally according to the yarn guides. This then makes it possible for the layers of the stitch-bonded fabric to be arranged freely and be made symmetrical in one working step. This process is advantageous to the characteristics of the composite as the "residual stresses resulting from asymmetric alignment of the layers are avoided, [while] the tensile strength and the impact strength of the composite are improved".

Needle Shift

Needle shift technique is when "Both outer warp layers [are] secured in one procedure by incorporating a shift of the needle bar during the stitching process, creating endless possibilities for the arrangement and patterns in the stitch-bonding process.

Advantages

Producing textiles through the warp knitting process has the following advantages

- higher productivity rates than weaving
- variety of fabric constructions
- large working widths
- low stress rate on the yarn that allows for use of fibers such as glass, aramide and carbon
- the creation of three-dimensional structures that can be knitted on double needle bar raschels

Applications

Warp knitted fabrics have several industrial uses, including producing mosquito netting, tulle fabrics, sports wear, shoe fabric, fabrics for printing and advertising, coating substrates and laminating backgrounds.

Research is also being conducted into the use of warp knitted fabrics for industrial applications (for example, to reinforce concrete), and for the production of biotextiles.

Warp Knitting and Biotextiles

The warp knitting process is also being used to create biotextiles. For example, a warp knitted polyester cardiac support device has been created to attempt to limit the growth of diseased hearts by being installed tightly around the diseased heart. Current research on animals "have confirmed that...the implantation of the device reverses the disease state, which makes this an alternative innovative therapy for patients who have side effects from traditional drug regimes".

Cable Knitting

Cable knitting is a style of knitting in which textures of crossing layers are achieved by permuting stitches. For example, given four stitches appearing on the needle in the order *ABCD*, one might cross the first two (in front of or behind) the next two, so that in subsequent rows those stitches appear in the new order *CDAB*.

A cable-knit piece of fabric

Methods

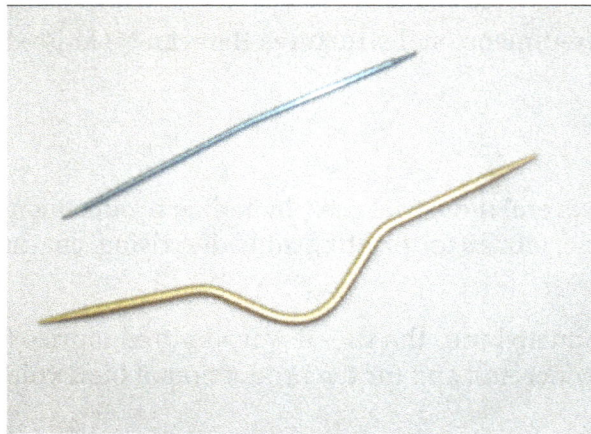

Two different styles of cable needles.

The stitches crossing behind are transferred to a small cable needle for storage while the stitches passing in front (or behind) are knitted. The former stitches are then transferred back to the original needle or knitted from the cable needle itself. Rather than use a cable needle, some knitters prefer to use a large safety pin or, for a single stitch, simply hold it in their fingers while knitting the other stitch(es). Cabling is typically done only when working on the right side of the fabric, i.e., every other row. This creates a *spacer row*, which helps the fabric to relax.

Cable knitting usually produces a fabric that is less flexible and more dense than typical knitting, having a much narrower gauge. This narrow gauge should be considered when changing from the cable stitch to another type of knitted fabric. If the number of stitches is not reduced, the second knitted fabric may flare out or pucker, due to its larger gauge. Thus, ribbed cuffs on an aran sweater may not contract around the wrist or waist, as would normally be expected. Conversely, stitches may need to be added to maintain the gauge when changing from another knitted fabric such as stocking to a cable pattern.

Cables are usually done in stocking stitch and surrounded with reverse stocking; this causes the cables to stand out against a receding background. But any combination will do; for example, a background seed stitch in the regions bounded by cables often looks striking. Another visually intriguing effect is meta-cabling, where the cable itself is made up of cables, such as a three-cable plait made of strands that are themselves 2-cable plaits. In such cases, the "inner" cables sometimes go their separate ways, forming beautiful, complex patterns such as the branches of a tree. Another interesting effect is to have one cable "pierce" another cable, rather than having it pass over or under the other.

Two cables should cross each other completely in a single row because making an intermediate crossing row of fewer stitches look good is very difficult. For example, where a pair of three-stitch-wide cables cross, all three stitches of one should cross over the three of the other cable.

Cable Braids

Cables are often used to make braid patterns. Usually, the cables themselves are with a knit stitch while the background is done in purl. As the number of cables increases, the number of crossing patterns increases, as described by the braid group. Various visual effects are also possible by shifting the center lines of the undulating cables, or by changing the space between the cables, making them denser or more open.

A one-cable serpentine is simply a cable that moves sinusoidally left and right as it progresses. Higher-order braids are often made with such serpentines crossing over and under each other.

A two-cable braid can look like a rope, if the cables always cross in the same way (e.g., left over right). Alternatively, it can look like two serpentines, one on top of the other.

A three-cable braid is usually a simple plait (as often seen styled in long hair), but can also be made to look like the links in a chain, or as three independent serpentines.

A four-cable braid allows for several crossing patterns.

The five-cable braid is sometimes called the Celtic princess braid, and is visually interesting because one side is cresting while the other side is in a trough. Thus, it has a shimmering quality, similar to a kris dagger.

The six-cable braid is called a Saxon braid, and looks square and solid. This is a large motif, often used as a centerpiece of an aran sweater or along the neckline and hemlines.

The seven-cable braid is rarely used, possibly because it is very wide.

It may be helpful to think of a cable pattern as a set of serpentine or wave-like cables, each one meandering around its own center line. A vast variety of cable patterns can be invented by changing the number of cables, the separations of their center lines, the amplitudes of their waves (i.e., how far they wander from their center line), the shape of the waves (e.g., sinusoidal versus triangular), and the relative position of the crests and troughs of each wave (e.g., is one wave cresting as another is crossing its center line?).

New cable patterns can also be inspired by pictures, scenes from nature, Celtic knotwork, and even the double helix of DNA.

Cable Lattices

In some cases, one can form a lattice of cables, a kind of ribbing made of cables where the individual cable strands can be exchanged freely. A typical example is a set of parallel two-cable plaits in which, every so often, the two cables of each plait separate, going left and right and integrating themselves in the neighboring cables. In the process, the right-going cable of one plait crosses the left-going cable of its neighbor, forming an "X".

Cable Textures

Many patterns made with cables do not have a rope-like quality. For example, a deep honeycomb pattern can be made by adjacent serpentines, first touching the neighbor on the left then the neighbor on the right. Other common patterns include a "Y"-like shape (and its inverse) and a horseshoe crab pattern.

Aran Sweaters

Many consider cable knitting to reach its heights in the Aran sweater, which consists of panels of different cable patterns.

References

- Bartlett, Roxana (1998). Slip-Stitch Knitting: Color Pattern the Easy Way. Loveland, CO: Interweave Press. ISBN 978-1-883010-32-4.

- Hollingworth, Shelagh (1983). The Complete Book of Traditional Aran Knitting. St. Martin's Press. ISBN 978-0-312-15635-0.

- Games, Alex (2007). Balderdash & piffle : one sandwich short of a dog's dinner. London: BBC. ISBN 978-1-84607-235-2.

- Brewer, John; Porter, Roy, eds. (1994). Consumption and the World of Goods. London: Routledge. pp. 232–233. ISBN 978-0-415-11478-3. LCCN 93180136.

- Silva, Marcos (2008). MALHARIA - BASES DE FUNDAMENTAÇÃO. Universidade Federal do Rio Grande do Norte. p. 2. Retrieved 22 December 2014.

6

E-Textiles: An Emerging Technology

The revolutionary field of e-textiles integrates technology into fabric by producing garments that have digital components embedded into them. This chapter introduces the reader to the topic of electronic textiles or smart textiles, its history and categories. The various applications of these textiles and its future have also been explored in this chapter.

E-textiles

E-textiles, also known as smart garments, smart clothing, electronic textiles, smart textiles, or smart fabrics, are fabrics that enable digital components (including small computers), and electronics to be embedded in them. Smart textiles are fabrics that have been developed with new technologies that provide added value to the wearer. Pailes-Friedman of the Pratt Institute states that "what makes smart fabrics revolutionary is that they have the ability to do many things that traditional fabrics cannot, including communicate, transform, conduct energy and even grow".

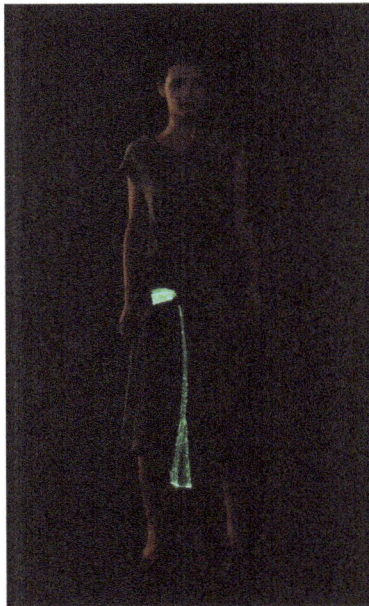

LEDs and fiber optics as part of women's fashion

Smart textiles can be broken into two different categories: aesthetic and performance enhancing. Aesthetic examples include everything from fabrics that light up to fabrics that can change color. Some of these fabrics gather energy from the environment by harnessing vibrations, sound or heat, reacting to this input. Then there are performance enhancing smart textiles, which will have a huge impact on the athletic, extreme sports and military industries. There are fabrics that help regulate body temperature, reduce wind resistance and control muscle vibration – all of which help improve athletic performance. Other fabrics have been developed for protective clothing to guard against extreme environmental hazards like radiation and the effects of space travel. The health and beauty industry is also taking advantage of these innovations, which range from drug-releasing medical textiles, to fabric with moisturizer, perfume, and anti-aging properties. Many smart clothing, wearable technology, and wearable computing projects involve the use of e-textiles.

Close-knit conductor (Image: Hong Kong Polytechnic University)

Electronic textiles are distinct from wearable computing because emphasis is placed on the seamless integration of textiles with electronic elements like microcontrollers, sensors, and actuators. Furthermore, e-textiles need not be wearable. For instance, e-textiles are also found in interior design.

The related field of fibretronics explores how electronic and computational functionality can be integrated into textile fibers.

A new report from Cientifica Research, Smart Textiles and Wearables: Markets, Applications and Technologies examines the markets for textile based wearable technologies, the companies producing them and the enabling technologies. This is creating a 4th industrial revolution for the textiles and fashion industry worth over $130 billion by 2025. The report identifies three distinct generations of textile wearable technologies:

1. First generation is where a sensor is attached to apparel and is the approach currently taken by major sportswear brands such as Adidas, Nike and Under Armour

2. Second generation products embed the sensor in the garment as demonstrated by products from Samsung, Alphabet, Ralph Lauren and Flex.

3. In third generation wearables the garment is the sensor and a growing number of companies including AdvanPro, Tamicare and BeBop sensors are making rapid progress in creating pressure, strain and temperature sensors.

Third generation wearables represent a significant opportunity for new and established textile companies to add significant value without having to directly compete with Apple, Samsung and Intel.

The report predicts that the key growth areas will be initially sports and wellbeing followed by medical applications for patient monitoring. Technical textiles, fashion and entertainment will also be significant applications with the total market expected to rise to over \$130 billion by 2025 with triple digit compound annual growth rates across many applications.

History

The basic materials needed to construct e-textiles, conductive threads and fabrics have been around for over 1000 years. In particular, artisans have been wrapping fine metal foils, most often gold and silver, around fabric threads for centuries. Many of Queen Elizabeth I's gowns, for example, are embroidered with gold-wrapped threads.

At the end of the 19th century, as people developed and grew accustomed to electric appliances, designers and engineers began to combine electricity with clothing and jewelry—developing a series of illuminated and motorized necklaces, hats, broaches and costumes. For example, in the late 1800s, a person could hire young women adorned in light-studded evening gowns from the Electric Girl Lighting Company to provide cocktail party entertainment.

In 1968, the Museum of Contemporary Craft in New York City held a ground-breaking exhibition called Body Covering that focused on the relationship between technology and apparel. The show featured astronauts' space suits along with clothing that could inflate and deflate, light up, and heat and cool itself. Particularly noteworthy in this collection was the work of Diana Dew, a designer who created a line of electronic fashion, including electroluminescent party dresses and belts that could sound alarm sirens.

In 1985, inventor Harry Wainwright created the first fully animated sweatshirt. The shirt consisted of fiber optics, leads, and a microprocessor to control individual frames of animation. The result was a full color cartoon displayed on the surface of the shirt. in 1995, Wainwright went on to invent the first machine enabling fiber optics to be machined into fabrics, the process needed for manufacturing enough for mass markets and, in 1997, hired a German machine designer, Herbert Selbach, from Selbach Machinery to produce the world's first CNC machine able to automatically implant fiber optics into any flexible material. Receiving the first of a dozen patents based on LED/Optic displays and machinery in 1989, the first CNC machines went into production in 1998 beginning with the production of animated coats for Disney Parks in 1998. The first ECG bio-physical display jackets employing LED/optic displays were created by Wainwright and David Bychkov, the CEO of Exmovere at the time in 2005 using GSR sensors in a watch connected via Bluetooth to the embedded machine washable display in a denim jacket and were demonstrated at the Smart Fabrics Conference held in Washington, D.C. May 7, 2007. Additional smart fabric technologies were unveiled by Wainwright at two Flextech Flexible Display conferences held in Phoenix, AZ, showing infrared digital displays machine-embedded into fabrics for IFF (Identification of Friend or Foe) which were submitted to BAE Systems for evaluation in 2006 and won an "Honorable Mention" award from NASA in 2010 on their Tech Briefs, "Design the Future" contest. MIT personnel purchased several fully animated coats for their researchers to wear at their demonstrations in 1999 to bring attention to their "Wearable Computer" research. Wainwright was commissioned to speak at the Textile and Colorists Conference in Melbourne, Australia on June 5, 2012 where he was requested to demonstrate his fabric creations that change color using any smart phone, indicate callers on mobile phones without a digital display, and contain WIFI security features that protect purses and personal items from theft.

In the mid 1990s a team of MIT researchers led by Steve Mann, Thad Starner, and Sandy Pentland began to develop what they termed wearable computers. These devices consisted of traditional computer hardware attached to and carried on the body. In response to technical, social, and design challenges faced by these researchers, another group at MIT, that included Maggie Orth and Rehmi Post, began to explore how such devices might be more gracefully integrated into clothing and other soft substrates. Among other developments, this team explored integrating digital electronics with conductive fabrics and developed a method for embroidering electronic circuits. One of the first commercially available wearable Arduino based microcontrollers, called the Lilypad Arduino, was also created at the MIT Media Lab by Leah Buechley.

Fashion houses like CuteCircuit are utilizing e-textiles for their haute couture collections and specialty projects. CuteCircuit's Hug Shirt allows the user to send electronic hugs through sensors within the garment.

Overview

The field of e-textiles can be divided into two main categories:

- E-textiles with classical electronic devices such as conductors, integrated circuits, LEDs, and conventional batteries embedded into garments.

- E-textiles with electronics integrated directly into the textile substrates. This can include either passive electronics such as conductors and resistors or active components like transistors, diodes, and solar cells.

Most research and commercial e-textile projects are hybrids where electronic components embedded in the textile are connected to classical electronic devices or components. Some examples are touch buttons that are constructed completely in textile forms by using conducting textile weaves, which are then connected to devices such as music players or LEDs that are mounted on woven conducting fiber networks to form displays.

Printed sensors for both physiological and environmental monitoring have been integrated into textiles including cotton, Gore-Tex, and neoprene.

Fibretronics

Just as in classical electronics, the construction of electronic capabilities on textile fibers requires the use of conducting and semi-conducting materials such as a conductive textile. There are a number of commercial fibers today that include metallic fibers mixed with textile fibers to form conducting fibers that can be woven or sewn. However, because both metals and classical semiconductors are stiff material, they are not very suitable for textile fiber applications, since fibers are subjected to much stretch and bending during use.

One of the most important issues of e-textiles is that the fibers should be washable. Electrical components would thus need to be insulated during washing to prevent damage.

A new class of electronic materials that are more suitable for e-textiles is the class of organic electronics materials, because they can be conducting, as well as semiconducting, and designed as inks and plastics.

Some of the most advanced functions that have been demonstrated in the lab include:

- Organic fiber transistors: the first textile fiber transistor that is completely compatible with textile manufacturing and that contains no metals at all.

- Organic solar cells on fibers

Uses

- Health monitoring of vital signs of the wearer such as heart rate, respiration rate, temperature, activity, and posture.

- Sports training data acquisition

- Monitoring personnel handling hazardous materials

- Tracking the position and status of soldiers in action

- Military app-Soldier's bullet proof kevlar vest,if the wearer is shot,the material can sense the bullet's impact and send a radio message back to base

- Monitoring pilot or truck driver fatigue

- Innovative Fashion(Wearable tech)

- Regain sensory perception that was previously lost by accident or birth.

References

- Smart Textiles and Wearables - Markets, Applications and Technologies. Innovation in Textiles (Report). September 7, 2016.

- Gaddis, Rebecca (May 7, 2014). "What Is The Future Of Fabric? These Smart Textiles Will Blow Your Mind". Forbes. Retrieved 2015-10-16.

- Cherenack, Kunigunde; Pieterson, Liesbeth van (2012-11-01). "Smart textiles: Challenges and opportunities". Journal of Applied Physics (published 7 November 2012). 112 (9): 091301. doi:10.1063/1.4742728. ISSN 0021-8979.

7

Uses of Textile

Textiles find use and application in various fields depending on their properties like finish, texture, abilities (fire retardancy, water- proofing, thermal insulation etc.) stretch and the like. This chapter aims at providing the reader a thorough understanding of the various types of specialized fabrics like Gore-Tex, geotextile, microfiber and spandex.

Clothing

Clothing in history, showing (from top) Egyptians, Ancient Greeks, Romans, Byzantines, Franks, and 13th through 15th century Europeans.

Clothing (also called clothes) is manufactured fiber and textile material worn on the body. The wearing of clothing is mostly restricted to humans and is a feature of nearly all human society. The amount and type of clothing one wears depend on physical requirements and local culture. Cultures regard some clothing types as gender-specific.

Origin of Clothing

There is no easy way to determine when clothing was first developed, but some information has been inferred by studying lice. The body louse specifically lives in clothing, and diverged from head lice about 107,000 years ago, suggesting that clothing existed at that time. Another theory is that modern humans are the only survivors of several species of primates who may have worn clothes and that clothing may have been used as long ago as 650 thousand years ago. Additional louse-based estimates put the introduction of clothing at around 42,000–72,000 BP. Other evidence suggests that humans may have begun wearing clothing as far back as 100,000 to 500,000 years ago.

Functions

Physically, clothing serves many purposes: it can serve as protection from the elements, and can enhance safety during hazardous activities such as hiking and cooking. It protects the wearer from rough surfaces, rash-causing plants, insect bites, splinters, thorns and prickles by providing a barrier between the skin and the environment. Clothes can insulate against cold or hot conditions. Further, they can provide a hygienic barrier, keeping infectious and toxic materials away from the body. Clothing also provides protection from harmful UV radiation. The most obvious function of clothing is to improve the comfort of the wearer, by protecting the wearer from the elements. In hot climates, clothing provides protection from sunburn or wind damage, while in cold climates its thermal insulation properties are generally more important. Shelter usually reduces the functional need for clothing. For example, coats, hats, gloves, and other superficial layers are normally removed when entering a warm home, particularly if one is residing or sleeping there. Similarly, clothing has seasonal and regional aspects, so that thinner materials and fewer layers of clothing are generally worn in warmer seasons and regions than in colder ones.

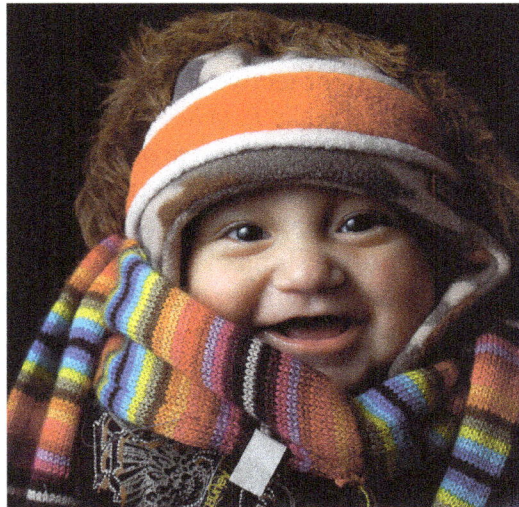

A baby wearing many items of winter clothing: headband, cap, fur-lined coat, shawl and sweater

Clothing performs a range of social and cultural functions, such as individual, occupational and sexual differentiation, and social status. In many societies, norms about clothing reflect standards of modesty, religion, gender, and social status. Clothing may also function as a form of adornment and an expression of personal taste or style.

Clothing can and has in history been made from a very wide variety of materials. Materials have ranged from leather and furs, to woven materials, to elaborate and exotic natural and synthetic fabrics. Not all body coverings are regarded as clothing. Articles carried rather than worn (such as purses), worn on a single part of the body and easily removed (scarves), worn purely for adornment (jewelry), or those that serve a function other than protection (eyeglasses), are normally considered accessories rather than clothing, as are footwear and hats.

Clothing protects against many things that might injure the uncovered human body. Clothes protect people from the elements, including rain, snow, wind, and other weather, as well as from the sun. However, clothing that is too sheer, thin, small, tight, etc., offers less protection. Clothes also reduce risk during activities such as work or sport. Some clothing protects from specific environmental hazards, such as insects, noxious chemicals, weather, weapons, and contact with abrasive substances. Conversely, clothing may protect the environment from the clothing *wearer*, as with doctors wearing medical scrubs.

Humans have shown extreme inventiveness in devising clothing solutions to environmental hazards. Examples include: space suits, air conditioned clothing, armor, diving suits, swimsuits, bee-keeper gear, motorcycle leathers, high-visibility clothing, and other pieces of protective clothing. Meanwhile, the distinction between clothing and protective equipment is not always clear-cut—since clothes designed to be fashionable often have protective value and clothes designed for function often consider fashion in their design. Wearing clothes also has social implications. They cover parts of the body that social norms require to be covered, act as a form of adornment, and serve other social purposes.

Scholarship

Although dissertations on clothing and its function appear from the 19th century as colonising countries dealt with new environments, concerted scientific research into psycho-social, physiological and other functions of clothing (e.g. protective, cartage) occurred in the first half of the 20th century, with publications such as J. C. Flügel's *Psychology of Clothes* in 1930, and Newburgh's seminal *Physiology of Heat Regulation and The Science of Clothing* in 1949. By 1968, the field of environmental physiology had advanced and expanded significantly, but the science of clothing in relation to environmental physiology had changed little. While considerable research has since occurred and the knowledge-base has grown significantly, the main concepts remain unchanged, and indeed Newburgh's book is still cited by contemporary authors, including those attempting to develop thermoregulatory models of clothing development.

Cultural Aspects

Gender Differentiation

In most cultures, gender differentiation of clothing is considered appropriate for men and women. The differences are in styles, colors and fabrics.

In Western societies, skirts, dresses and high-heeled shoes are usually seen as women's clothing, while neckties are usually seen as men's clothing. Trousers were once seen as exclusively male clothing, but are nowadays worn by both genders. Male clothes are often more practical (that is, they can function well under a wide variety of situations), but a wider range of clothing styles are available for females. Males are typically allowed to bare their chests in a greater variety of public places. It is generally acceptable for a woman to wear traditionally male clothing, while the converse is unusual.

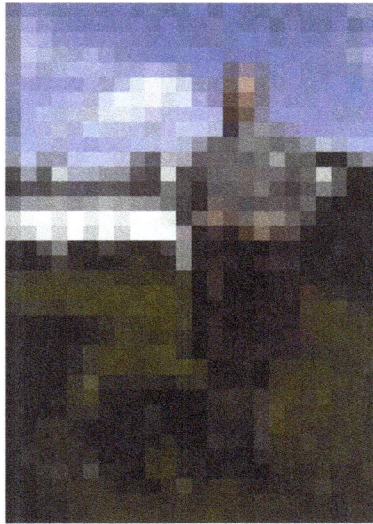

Former 3rd Duke of Fife wearing a traditional Scottish kilt. (1984)

Former US Secretary of State Condoleezza Rice and Turkish President Abdullah Gül both wearing Western-style business suits.

In some cultures, sumptuary laws regulate what men and women are required to wear. Islam requires women to wear more modest forms of attire, usually hijab. What qualifies as "modest" varies in different Muslim societies. However, women are usually required to cover more of their bodies than men are. Articles of clothing Muslim women wear for modesty range from the head-scarf to the burqa.

Men may sometimes choose to wear men's skirts such as togas or kilts, especially on ceremonial occasions. Such garments were (in previous times) often worn as normal daily clothing by men.

Social Status

In some societies, clothing may be used to indicate rank or status. In ancient Rome, for example, only senators could wear garments dyed with Tyrian purple. In traditional Hawaiian society, only high-ranking chiefs could wear feather cloaks and palaoa, or carved whale teeth. Under the Travancore Kingdom of Kerala, (India), lower caste women had to pay a tax for the right to cover their upper body. In China, before establishment of the republic, only the emperor could wear yellow. History provides many examples of elaborate sumptuary laws that regulated what people could wear. In societies without such laws, which includes most modern societies, social status is instead signaled by the purchase of rare or luxury items that are limited by cost to those with wealth or status. In addition, peer pressure influences clothing choice.

A barong Tagalog made from piña fiber.

Alim Khan's bemedaled robe sends a social message about his wealth, status, and power

Religion

Religious clothing might be considered a special case of occupational clothing. Sometimes it is worn only during the performance of religious ceremonies. However, it may also be worn everyday as a marker for special religious status.

Nicolas Trigault, a Flemish Jesuit, in Ming-style Confucian scholar costume, by Peter Paul Rubens. This drawing shows a fusion between West and East also Christianity and Confucianism.

Muslim men traditionally wear white robes and a cap during prayers

For example, Jains and Muslim men wear unstitched cloth pieces when performing religious ceremonies. The unstitched cloth signifies unified and complete devotion to the task at hand, with no digression. Sikhs wear a turban as it is a part of their religion.

The cleanliness of religious dresses in Eastern religions like Hinduism, Sikhism, Buddhism, Islam and Jainism is of paramount importance, since it indicates purity.

Clothing figures prominently in the Bible where it appears in numerous contexts, the more prominent ones being: the story of Adam and Eve who made coverings for themselves out of fig leaves, Joseph's cloak, Judah and Tamar, Mordecai and Esther. Furthermore, the priests officiating in the Temple had very specific garments, the lack of which made one liable to death.

In Islamic traditions, women are required to wear long, loose, opaque outer dress when stepping out of the home. This dress code was democratic (for all women regardless of status) and for protection from the scorching sun. The Quran says this about husbands and wives: "…They are clothing/covering (Libaas) for you; and you for them" (chapter 2:187).

Jewish ritual also requires rending of one's upper garment as a sign of mourning. This practice is found in the Bible when Jacob hears of the apparent death of his son Joseph.

Origin and History

First Recorded Use

According to archaeologists and anthropologists, the earliest clothing likely consisted of fur, leather, leaves, or grass that were draped, wrapped, or tied around the body. Knowledge of such clothing remains inferential, since clothing materials deteriorate quickly compared to stone, bone, shell and metal artifacts. Archeologists have identified very early sewing needles of bone and ivory from about 30,000 BC, found near Kostenki, Russia in 1988. Dyed flax fibers that could have been used in clothing have been found in a prehistoric cave in the Republic of Georgia that date back to 36,000 BP.

Scientists are still debating when people started wearing clothes. Ralf Kittler, Manfred Kayser and Mark Stoneking, anthropologists at the Max Planck Institute for Evolutionary Anthropology, have conducted a genetic analysis of human body lice that suggests clothing originated quite recently, around 170,000 years ago. Body lice is an indicator of clothes-wearing, since most humans have sparse body hair, and lice thus require human clothing to survive. Their research suggests the invention of clothing may have coincided with the northward migration of modern *Homo sapiens* away from the warm climate of Africa, thought to have begun between 50,000 and 100,000 years ago. However, a second group of researchers using similar genetic methods estimate that clothing originated around 540,000 years ago (Reed et al. 2004. PLoS Biology 2(11): e340). For now, the date of the origin of clothing remains unresolved.

Making Clothing

Some human cultures, such as the various people of the Arctic Circle, traditionally make their clothing entirely of prepared and decorated furs and skins. Other cultures supplemented or replaced leather and skins with cloth: woven, knitted, or twined from various animal and vegetable fibers.

Although modern consumers may take the production of clothing for granted, making fabric by hand is a tedious and labor-intensive process. The textile industry was the first to be mechanized – with the powered loom – during the Industrial Revolution.

Different cultures have evolved various ways of creating clothes out of cloth. One approach simply involves draping the cloth. Many people wore, and still wear, garments consisting of rectangles of cloth wrapped to fit – for example, the dhoti for men and the sari for women in the Indian subcontinent, the Scottish kilt or the Javanese sarong. The clothes may simply be tied up, as is the case of the first two garments; or pins or belts hold the garments in place, as in the case of the latter two. The precious cloth remains uncut, and people of various sizes or the same person at different sizes can wear the garment.

Another approach involves cutting and sewing the cloth, but using every bit of the cloth rectangle in constructing the clothing. The tailor may cut triangular pieces from one corner of the cloth, and then add them elsewhere as gussets. Traditional European patterns for men's shirts and women's chemises take this approach.

Modern European fashion treats cloth much less conservatively, typically cutting in such a way as to leave various odd-shaped cloth remnants. Industrial sewing operations sell these as waste; home sewers may turn them into quilts.

In the thousands of years that humans have spent constructing clothing, they have created an astonishing array of styles, many of which have been reconstructed from surviving garments, photos, paintings, mosaics, etc., as well as from written descriptions. Costume history serves as a source of inspiration to current fashion designers, as well as a topic of professional interest to costumers constructing for plays, films, television, and historical reenactment.

Contemporary Clothing

Western Dress Code

The Western dress code has changed over the past 500+ years. The mechanization of the textile industry made many varieties of cloth widely available at affordable prices. Styles have changed, and the availability of synthetic fabrics has changed the definition of "stylish". In the latter half of the 20th century, blue jeans became very popular, and are now worn to events that normally demand formal attire. Activewear has also become a large and growing market.

The licensing of designer names was pioneered by designers like Pierre Cardin in the 1960s and has been a common practice within the fashion industry from about the 1970s.

Spread of Western Styles

Ottoman representation of European dress circa 1780.

By the early years of the 21st century, western clothing styles had, to some extent, become international styles. This process began hundreds of years earlier, during the periods of European colonialism. The process of cultural dissemination has perpetuated over the centuries as Western media corporations have penetrated markets throughout the world, spreading Western culture and styles. Fast fashion clothing has also become a global phenomenon. These garments are less expensive, mass-produced Western clothing. Donated used clothing from Western countries are also delivered to people in poor countries by charity organizations.

Ethnic and Cultural Heritage

People may wear ethnic or national dress on special occasions or in certain roles or occupations. For example, most Korean men and women have adopted Western-style dress for daily wear, but still wear traditional hanboks on special occasions, like weddings and cultural holidays. Items of Western dress may also appear worn or accessorized in distinctive, non-Western ways. A Tongan man may combine a used T-shirt with a Tongan wrapped skirt, or tupenu.

Sport and Activity

Fashion shows are often the source of the latest style and trends in clothing fashions.

Most sports and physical activities are practiced wearing special clothing, for practical, comfort or safety reasons. Common sportswear garments include shorts, T-shirts, tennis shirts, leotards, tracksuits, and trainers. Specialized garments include wet suits (for swimming, diving or surfing), salopettes (for skiing) and leotards (for gymnastics). Also, spandex materials are often used as base layers to soak up sweat. Spandex is also preferable for active sports that require form fitting garments, such as volleyball, wrestling, track & field, dance, gymnastics and swimming.

Fashion

There exists a diverse range of styles in fashion, varying by geography, exposure to modern media, economic conditions, and ranging from expensive haute couture to traditional garb, to thrift store grunge.

Future Trends

The world of clothing is always changing, as new cultural influences meet technological innovations. Researchers in scientific labs have been developing prototypes for fabrics that can serve functional purposes well beyond their traditional roles, for example, clothes that can automatically adjust their temperature, repel bullets, project images, and generate electricity. Some practical advances already available to consumers are bullet-resistant garments made with kevlar and stain-resistant fabrics that are coated with chemical mixtures that reduce the absorption of liquids.

Political Issues

Working Conditions

Safety garb for women workers in Los Angeles, California, ca. 1943. The uniform at the left, complete with the plastic "bra" on the right, was designed to prevent occupational accidents among female war workers.

Though mechanization transformed most aspects of human industry by the mid-20th century, garment workers have continued to labor under challenging conditions that demand repetitive manual labor. Mass-produced clothing is often made in what are considered by some to be sweatshops, typified by long work hours, lack of benefits, and lack of worker representation. While most examples of such conditions are found in developing countries, clothes made in industrialized nations may also be manufactured similarly, often staffed by undocumented immigrants.

Coalitions of NGOs, designers (Katharine Hamnett, American Apparel, Veja, Quiksilver, eVocal, Edun,…) and campaign groups like the Clean Clothes Campaign (CCC) and the Institute for Global Labour and Human Rights as well as textile and clothing trade unions have sought to improve these conditions as much as possible by sponsoring awareness-raising events, which draw the attention of both the media and the general public to the workers.

Outsourcing production to low wage countries like Bangladesh, China, India and Sri Lanka became possible when the Multi Fibre Agreement (MFA) was abolished. The MFA, which placed quotas on textiles imports, was deemed a protectionist measure. Globalization is often quoted as the single most contributing factor to the poor working conditions of garment workers. Although many countries recognize treaties like the International Labour Organization, which attempt to set standards for worker safety and rights, many countries have made exceptions to certain parts of the treaties or failed to thoroughly enforce them. India for example has not ratified sections 87 and 92 of the treaty.

Despite the strong reactions that "sweatshops" evoked among critics of globalization, the production of textiles has functioned as a consistent industry for developing nations providing work and wages, whether construed as exploitative or not, to many thousands of people.

Fur

The use of animal fur in clothing dates to prehistoric times. It is currently associated in developed countries with expensive, designer clothing, although fur is still used by indigenous people in arctic zones and higher elevations for its warmth and protection. Once uncontroversial, it has recently been the focus of campaigns on the grounds that campaigners consider it cruel and unnecessary. PETA, along with other animal rights and animal liberation groups have called attention to fur farming and other practices they consider cruel.

Life Cycle

Clothing Maintenance

Clothing suffers assault both from within and without. The human body sheds skin cells and body oils, and exudes sweat, urine, and feces. From the outside, sun damage, moisture, abrasion and dirt assault garments. Fleas and lice can hide in seams. Worn clothing, if not cleaned and refurbished, itches, looks scruffy, and loses functionality (as when buttons fall off, seams come undone, fabrics thin or tear, and zippers fail).

In some cases, people wear an item of clothing until it falls apart. Cleaning leather presents difficulties, and bark cloth (tapa) cannot be washed without dissolving it. Owners may patch tears and rips, and brush off surface dirt, but old leather and bark clothing always look *old*.

But most clothing consists of cloth, and most cloth can be laundered and mended (patching, darning, but compare felt).

Laundry, Ironing, Storage

Humans have developed many specialized methods for laundering, ranging from early methods of pounding clothes against rocks in running streams, to the latest in electronic washing machines and dry cleaning (dissolving dirt in solvents other than water). Hot water washing (boiling), chemical cleaning and ironing are all traditional methods of sterilizing fabrics for hygiene purposes.

Many kinds of clothing are designed to be ironed before they are worn to remove wrinkles. Most modern formal and semi-formal clothing is in this category (for example, dress shirts and suits). Ironed clothes are believed to look clean, fresh, and neat. Much contemporary casual clothing is made of knit materials that do not readily wrinkle, and do not require ironing. Some clothing is permanent press, having been treated with a coating (such as polytetrafluoroethylene) that suppresses wrinkles and creates a smooth appearance without ironing.

Once clothes have been laundered and possibly ironed, they are usually hung on clothes hangers or folded, to keep them fresh until they are worn. Clothes are folded to allow them to be stored compactly, to prevent creasing, to preserve creases or to present them in a more pleasing manner, for instance when they are put on sale in stores.

Non-iron

A resin used for making non-wrinkle shirts releases formaldehyde, which could cause contact dermatitis for some people; no disclosure requirements exist, and in 2008 the U.S. Government Accountability Office tested formaldehyde in clothing and found that generally the highest levels were in non-wrinkle shirts and pants. In 1999, a study of the effect of washing on the formaldehyde levels found that after 6 months after washing, 7 of 27 shirts had levels in excess of 75 ppm, which is a safe limit for direct skin exposure.

Mending

In past times, mending was an art. A meticulous tailor or seamstress could mend rips with thread raveled from hems and seam edges so skillfully that the tear was practically invisible. When the raw material – cloth – was worth more than labor, it made sense to expend labor in saving it. Today clothing is considered a consumable item. Mass-manufactured clothing is less expensive than the labor required to repair it. Many people buy a new piece of clothing rather than spend time mending. The thrifty still replace zippers and buttons and sew up ripped hems.

Recycling

Used, unwearable clothing can be used for quilts, rags, rugs, bandages, and many other household uses. It can also be recycled into paper. In Western societies, used clothing is often thrown out or donated to charity (such as through a clothing bin). It is also sold to consignment shops, dress agencies, flea markets, and in online auctions. Used clothing is also often collected on an industrial scale to be sorted and shipped for re-use in poorer countries.

There are many concerns about the life cycle of synthetics, which come primarily from petrochemicals. Unlike natural fibers, their source is not renewable and they are not biodegradable.

Spandex

Spandex, Lycra or elastane is a synthetic fiber known for its exceptional elasticity. It is stronger and more durable than natural rubber. It is a polyester-polyurethane copolymer that was invented in 1958 by chemist Joseph Shivers at DuPont's Benger Laboratory in Waynesboro, Virginia. When introduced in 1962, it revolutionized many areas of the clothing industry.

American volleyball player Cynthia Barboza wearing spandex shorts

The name "spandex" is an anagram of the word "expands". It is the preferred name in North America; in continental Europe it is referred to by variants of "elastane", i.e. *élasthanne* (France), *Elastan* (Germany), *elastano* (Spain), *elastam* (Italy) and *elastaan* (Netherlands), and is known in the UK, Ireland, Portugal, Brazil, Argentina, Australia, New Zealand and Israel primarily as Lycra. Brand names for spandex include Lycra (made by Koch subsidiary Invista, previously a part of DuPont), Elaspan (also Invista), Acepora (Taekwang), Creora (Hyosung), INVIYA (Indorama Corporation), ROICA and Dorlastan (Asahi Kasei), Linel (Fillattice), and ESPA (Toyobo).

Production

Spandex fibers are produced in four different ways: melt extrusion, reaction spinning, solution dry spinning, and solution wet spinning. All of these methods include the initial step of reacting monomers to produce a prepolymer. Once the prepolymer is formed, it is reacted further in various ways and drawn out to make the fibers.

Spandex fiber

The solution dry spinning method is used to produce over 94.5% of the world's spandex fibers, and the process has five steps:

1. The first step is to produce the prepolymer. This is done by mixing a macroglycol with a diisocyanate monomer. The two compounds are mixed in a reaction vessel to produce a prepolymer. A typical ratio of glycol to diisocyanate is 1:2.

2. The prepolymer is further reacted with an equal amount of diamine. This reaction is known as *chain extension reaction*. The resulting solution is diluted with a solvent (DMAc) to produce the spinning solution. The solvent helps make the solution thinner and more easily handled, and then it can be pumped into the fiber production cell.

3. The spinning solution is pumped into a cylindrical spinning cell where it is cured and converted into fibers. In this cell, the polymer solution is forced through a metal plate called a spinneret. This causes the solution to be aligned in strands of liquid polymer. As the strands pass through the cell, they are heated in the presence of a nitrogen and solvent gas. This process causes the liquid polymer to react chemically and form solid strands.

4. As the fibers exit the cell, an amount of solid strands are bundled together to produce the desired thickness. Each fiber of spandex is made up of many smaller individual fibers that adhere to one another due to the natural stickiness of their surface.

5. The resulting fibers are then treated with a finishing agent which can be magnesium stearate or another polymer. This treatment prevents the fibers' sticking together and aids in textile manufacture. The fibers are then transferred through a series of rollers onto a spool.

Major Spandex Fiber Uses

Cyclist wearing a pair of spandex shorts and a cycling jersey

Because of its elasticity and strength (stretching up to five times its length), spandex has been incorporated into a wide range of garments, especially in skin-tight garments. A benefit of spandex is its significant strength and elasticity and its ability to return to the original shape after stretching and faster drying than ordinary fabrics.

Woman wearing spandex leggings

Wrestlers wearing spandex

The types of garments which incorporate spandex include:

- Apparel and clothing articles where stretch is desired, generally for comfort and fit, such as:
    - activewear
    - athletic, aerobic, and exercise apparel
    - belts
    - bra straps and side panels
    - competitive swimwear
    - cycling jerseys and shorts
    - dance belts worn by male ballet dancers and others
    - gloves
    - hosiery
    - leggings
    - netball bodysuits
    - orthopaedic braces
    - rowing unisuits
    - cross country race suits
    - ski pants
    - skinny jeans
    - slacks
    - miniskirts
    - socks and tights
    - swimsuits/bathing suits
    - underwear
    - wetsuits
    - zentai
    - Triathlon suits
- Compression garments such as:
    - foundation garments
    - motion capture suits

- Shaped garments such as:
    - bra cups
    - support hose
    - surgical hose
    - superhero outfits
    - women's volleyball shorts
    - wrestling singlets
- Home furnishings, such as microbead pillows

For clothing, spandex is usually mixed with cotton or polyester, and accounts for a small percentage of the final fabric, which therefore retains most of the look and feel of the other fibers. In North America it is rare in men's clothing, but prevalent in women's. An estimated 80% of clothing sold in the United States contained spandex in 2010.

Gore-tex

Gore-Tex is a waterproof, breathable fabric membrane and registered trademark of W. L. Gore and Associates. Invented in 1969, Gore-Tex is able to repel liquid water while allowing water vapor to pass through, and is designed to be a lightweight, waterproof fabric for all-weather use. It is composed of stretched polytetrafluoroethylene (PTFE), which is more commonly known as the generic trademark Teflon.

History

Prior to Gore-Tex, stretched polytetrafluoroethylene (PTFE) tape was produced in 1966 by John W. Cropper of New Zealand. He had developed and constructed a machine for this use. However, Cropper chose to keep the process of creating expanded PTFE as a closely held trade secret and as such, it remained unpublished. As a public patent had not been filed, the new form of the material could not be legally recognised.

Gore-Tex was co-invented by Wilbert L. Gore and Gore's son, Robert W. Gore. In 1969, Bob Gore stretched heated rods of PTFE and created expanded polytetrafluoroethylene (ePTFE). His discovery of the right conditions for stretching PTFE was a happy accident, born partly of frustration. Instead of slowly stretching the heated material, he applied a sudden, accelerating yank. The solid PTFE unexpectedly stretched about 800%, forming a microporous structure that was about 70% air. It was introduced to the public under the trademark Gore-Tex. Bob Gore promptly applied for and obtained the following patents:

- U.S. Patent 3,953,566, issued April 27, 1976, for a porous form of polytetrafluoroethylene with a micro-structure characterized by nodes interconnected by fibrils
- U.S. Patent 4,187,390, issued February 5, 1980

- U.S. Patent 4,194,041 on March 18, 1980 for a "waterproof laminate", together with Samuel Allen

In the 1970s Garlock, Inc. allegedly infringed Gore's patents and was sued by Gore in the Federal District Court of Ohio. The District Court held Gore's patents to be invalid after a "bitterly contested case" that "involved over two years of discovery, five weeks of trial, the testimony of 35 witnesses (19 live, 16 by deposition), and over 300 exhibits" (quoting the Federal Circuit). On appeal, however, the Federal Circuit disagreed in the famous case of *Gore v. Garlock*, reversing the lower court's decision on the ground, as well as others, that Cropper forfeited any superior claim to the invention by virtue of having concealed the process for making ePTFE from the public, thereby establishing Gore as the legal inventor.

Gore-Tex is used in products manufactured by Patagonia, L.L. Bean, Oakley, Inc., Galvin Green, Marmot, Vasque, Arc'teryx, Haglöfs and The North Face among others.

Since the expiration of the main Gore-Tex patent, several other products have come to market with similar characteristics that use similar technology. As a cheaper alternative to the more expensive membranes, there are also waterproof/breathable coatings which are less durable.

For his invention, Robert W. Gore was inducted into the U.S. National Inventors Hall of Fame in 2006.

Manufacture

PTFE is made using an emulsion polymerization process that utilizes the carcinogenic fluorosurfactant PFOA, a persistent environmental contaminant. In 2013, Gore eliminated the use of PFOAs in the manufacture of its weatherproof functional fabrics.

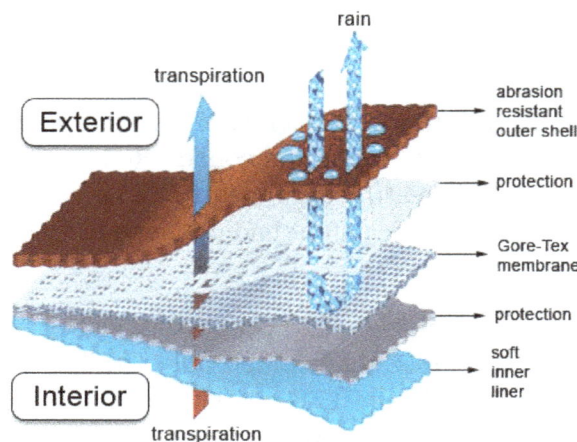

Schematic of a composite Gore-Tex fabric for outdoor clothing

Design

Gore-Tex materials are typically based on thermo-mechanically expanded PTFE and other fluoropolymer products. They are used in a wide variety of applications such as high-performance fabrics, medical implants, filter media, insulation for wires and cables, gaskets, and sealants. However, Gore-Tex fabric is best known for its use in protective, yet breathable, rainwear.

Waterproof running shoes made with Gore-Tex, offered by different manufacturers, exhibited to running club meeting at local shoe store.

The simplest sort of rain wear is a two layer sandwich. The outer layer is typically nylon or polyester and provides strength. The inner one is polyurethane (abbreviated: PU), and provides water resistance, at the cost of breathability.

Early Gore-Tex fabric replaced the inner layer of PU with a thin, porous fluoropolymer membrane (Teflon) coating that is bonded to a fabric. This membrane had about 9 billion pores per square inch (around 1.4 billion pores per square centimeter). Each pore is approximately 1/20,000 the size of a water droplet, making it impenetrable to liquid water while still allowing the more volatile water vapour molecules to pass through.

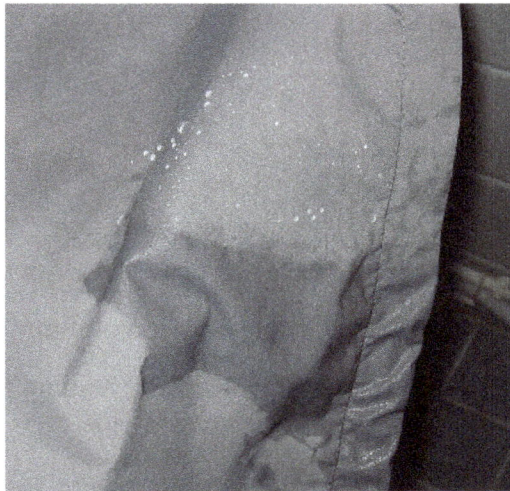

Effect of water repellent on a shell layer Gore-Tex jacket (Haglöfs Heli II).

Both wear and cleaning will reduce the performance of Gore-Tex clothes by wearing away the Durable Water Repellent (DWR) treatment on the surface of the fabric. The DWR prevents the face fabric from becoming wet and thus reducing breathability. However, the DWR is not responsible for the jacket being waterproof. This is a common misconception, so when the face fabric becomes soaked due to an absence of DWR, there is no breathability and the wearer's sweat will cause condensation to form inside the jacket. This may give the appearance that a jacket is leaking when it is not. The DWR can be reinvigorated by tumble drying the garment or ironing on a low setting.

Gore requires that all garments made from their material have taping over the seams, to eliminate leaks. Gore's sister product, Windstopper, is similar to Gore-Tex in being windproof and breathable, but has ability to stretch and is not waterproof. The Gore naming system does not imply specific technology or material but instead specific set of performance characteristics.

Other Uses

Gore-Tex is also used internally in medical applications, because it is nearly inert inside the body. In addition, the porosity of Gore-Tex permits the body's own tissue to grow through the material, integrating grafted material into the circulation system. Gore-Tex is used in a wide variety of medical applications, including sutures, vascular grafts, heart patches, and synthetic knee ligaments, which have saved thousands of lives.

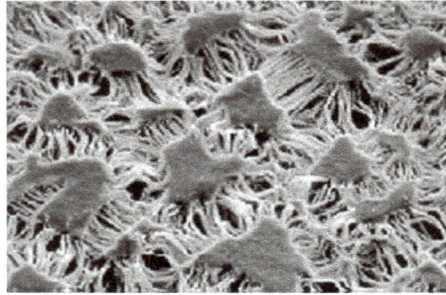

Gore-Tex membrane under an electron microscope

Gore-Tex Medical Devices Sample Kit, Chemical Heritage Foundation

Gore-Tex has been used for many years in the conservation of illuminated manuscripts.

Explosive sensors have been printed on Gore-Tex clothing leading to the sensitive voltammetric detection of nitroaromatic compounds.

The "Gore-Tex" brand name was formerly used for industrial and medical products.

Geotextile

Geotextiles are permeable fabrics which, when used in association with soil, have the ability to separate, filter, reinforce, protect, or drain. Typically made from polypropylene or polyester, geotextile fabrics come in three basic forms: woven (resembling mail bag sacking), needle punched (resembling felt), or heat bonded (resembling ironed felt).

Examples of geotextiles

Geotextile composites have been introduced and products such as geogrids and meshes have been developed. Overall, these materials are referred to as geosynthetics and each configuration—-geonets, geogrids, geotubes (such as TITANTubes) and others—-can yield benefits in geotechnical and environmental engineering design.

History

Geotextiles were originally intended to be an alternative to granular soil filters. The original, and still sometimes used, term for geotextiles is *filter fabrics*. Work originally began in the 1950s with R.J. Barrett using geotextiles behind precast concrete seawalls, under precast concrete erosion control blocks, beneath large stone riprap, and in other erosion control situations. He used different styles of woven monofilament fabrics, all characterized by a relatively high percentage open area (varying from 6 to 30%). He discussed the need for both adequate permeability and soil retention, along with adequate fabric strength and proper elongation and set the tone for geotextile use in filtration situations.

Applications

Geotextiles and related products have many applications and currently support many civil engineering applications including roads, airfields, railroads, embankments, retaining structures, reservoirs, canals, dams, bank protection, coastal engineering and construction site silt fences or geotube. Usually geotextiles are placed at the tension surface to strengthen the soil. Geotextiles are also used for sand dune armoring to protect upland coastal property from storm surge, wave action and flooding. A large sand-filled container (SFC) within the dune system prevents storm erosion from proceeding beyond the SFC. Using a sloped unit rather than a single tube eliminates damaging scour.

A silt fence on a construction site.

Erosion control manuals comment on the effectiveness of sloped, stepped shapes in mitigating shoreline erosion damage from storms. Geotextile sand-filled units provide a "soft" armoring solution for upland property protection. Geotextiles are used as matting to stabilize flow in stream channels and swales.

Geotextiles can improve soil strength at a lower cost than conventional soil nailing. In addition, geotextiles allow planting on steep slopes, further securing the slope.

Geotextiles have been used to protect the fossil hominid footprints of Laetoli in Tanzania from erosion, rain, and tree roots.

In building demolition, geotextile fabrics in combination with steel wire fencing can contain explosive debris.

Coir (coconut fiber) geotextiles are popular for erosion control, slope stabilization and bioengineering, due to the fabric's substantial mechanical strength. App. L.e Coir geotextiles last approximately 3 to 5 years depending on the fabric weight. The product degrades into humus, enriching the soil.

Design Considerations

To use geotextiles to reinforce a steep slope, two components have to be calculated:

- the tension required for equilibrium
- the appropriate layout of the geotextile reinforcement.

Microfiber

Microfiber or microfibre is synthetic fiber finer than one denier or decitex/thread. This is smaller than the diameter of a strand of silk (which is approximately one denier), which is itself about 1/5 the diameter of a human hair. The most common types of microfibers are made from polyesters, polyamides (e.g., nylon, Kevlar, Nomex, trogamide), or a conjugation of polyester, polyamide, and polypropylene (Prolen). Microfiber is used to make mats, knits, and weaves for apparel, upholstery, industrial filters, and cleaning products. The shape, size, and combinations of synthetic fibers are selected for specific characteristics, including softness, toughness, absorption, water repellency, electrostatics, and filtering capabilities.

Close-Up view of Microfiber / Microfibre Cloth

Microfiber cloth suitable for cleaning sensitive surfaces

History

Production of ultra-fine fibers (finer than 0.7 denier) dates back to the late 1950s, using melt-blown spinning and flash spinning techniques. However, only fine staples of random length could be manufactured and very few applications could be found. Experiments to produce ultra-fine fibers of a continuous filament type were made subsequently, the most promising of which were run in Japan during the 1960s by Dr. Miyoshi Okamoto, a scientist at Toray Industries. Okamoto's discoveries, together with those of Dr. Toyohiko Hikota, resulted in many industrial applications. Among these was Ultrasuede, one of the first successful synthetic microfibers, which found its way onto the market in the 1970s. Microfiber's use in the textile industry then expanded. Microfibers were first publicized in the early 1990s in Sweden and saw success as a product in Europe over the course of the decade.

Apparel

Clothing

Microfiber fabric is often used for athletic wear, such as cycling jerseys, because the microfiber material wicks moisture (perspiration) away from the body, keeping the wearer cool and dry. Microfiber is also very elastic, making it suitable for undergarments. However, the US Marine Corps banned synthetic fabrics for wear with uniforms while deployed to combat environments in 2006, due to instances where Marines' undergarments were melting under extreme heat caused by IED (improvised explosive device) blasts, causing more damage to the skin. They released a "fit for duty" version authorized earlier that same year.

Microfiber is also used to make tough, very soft-to-the-touch materials for general clothing use, often used in skirts and jackets. Microfiber can be made into Ultrasuede, an animal-free imitation suede leather-like product that is cheaper and easier to clean and sew than natural suede leather.

Microfiber fabric can be used for making bathrobes, jackets, swim trunks, and other clothing that can be worn for aquatic activities such as swimming.

Accessories

Microfiber is used to make many accessories that traditionally have been made from leather: wallets, handbags, backpacks, book covers, shoes, cell phone cases, and coin purses. Microfiber fabric is lightweight, durable, and somewhat water repellent, so it makes a good substitute.

Another advantage of fabric (compared to leather) is that fabric can be coated with various finishes or can be treated with antibacterial chemicals. Fabric can also be printed with various designs, embroidered with colored thread, or heat-embossed to create interesting textures.

Other Uses

Textiles for Cleaning

In cleaning products, microfiber can be 100% polyester, or a blend of polyester and polyamide (nylon). It can be both a woven product or a non woven product, the latter most often used in limited use or disposable cloths. In the highest-quality fabrics for cleaning applications, the fiber is split during the manufacturing process to produce multi-stranded fibres. A cross section of the split microfiber fabric under high magnification would look like an asterisk. The split fibres and the size of the individual filaments working in conjunction with the spaces between them make the cloths more effective than other fabrics for cleaning purposes. The structure traps and retains the dirt and also absorbs liquids.

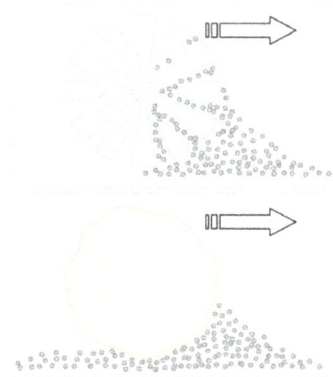

Cross sections: microfiber thread above, cotton thread below

Microfiber cloth for cleaning screens and lenses

Microfiber mop with velcro back for fastening on handle

Unlike cotton, microfiber leaves no lint, the exception being some micro suede blends, where the surface is mechanically processed to produce a soft plush feel.

For microfiber to be most effective as a cleaning product, especially for water-soluble soils and waxes, it should be a split microfiber. Non-split microfiber is little more than a very soft cloth. The main exception is for cloths used for facial cleansing and for the removal of skin oils, (sebum), sunscreens, and mosquito repellents from optical surfaces such as cameras, phones and eyeglasses where in higher-end proprietary woven, 100% polyester cloths using 2 μm filaments, will absorb these types of oils without smearing.

Microfiber that is used in non-sports-related clothing, furniture, and other applications isn't split because it isn't designed to be absorbent, just soft. When buying, microfiber may not be labelled to designate whether it is split. A quick way to determine if microfiber is, is to run the cloth lightly over the palm of the hand. A split microfiber will cling to any imperfections of the skin, which can be both heard and felt. Another way is to pour a small amount of water on a hard flat surface and try to push the water with the microfiber. If the water is pushed rather than being absorbed, it's not split microfiber..

Microfiber can be electrostatically-charged for special purposes like filtration.

Rags

Microfiber products used for consumer cleaning are generally constructed from split conjugated fibers of polyester and polyamide. Microfiber used for commercial cleaning products also includes many products constructed of 100% polyester microfiber. Fabrics made with microfibers are exceptionally soft and hold their shape well. When high-quality microfiber is combined with the right knitting process, it creates an extremely effective cleaning material. This material can hold up to eight times its weight in water. Microfiber products have exceptional ability to absorb oils, and are not hard enough to scratch even paintwork unless they have retained grit or hard particles from previous use.

Microfiber is widely used by car detailers to handle tasks such as removing wax from paintwork, quick detailing, cleaning interior, cleaning glass, and drying. Due to their fine fibers which leave no lint or dust, microfiber towels are used by car detailers and enthusiasts in a similar manner to a chamois leather.

Microfiber is used in many professional cleaning applications, for example in mops and cleaning cloths. Although microfiber mops cost more than non-microfiber mops, they may be more economical because they last longer and require less effort to use.

Microfiber textiles designed for cleaning clean on a microscopic scale. According to tests using microfiber materials to clean a surface leads to reducing the number of bacteria by 99%, whereas a conventional cleaning material reduces this number only by 33%. Microfiber cleaning tools also absorb fat and grease and their electrostatic properties give them a high dust-attracting power.

Microfiber cloths are used to clean photographic lenses as they absorb oily matter without being abrasive or leaving a residue, and are sold by major manufacturers such as Sinar, Nikon and Canon. Small microfiber cleaning cloths are commonly sold for cleaning computer screens and eyeglasses.

Microfiber is unsuitable for some cleaning applications as it accumulates dust, debris, and particles. Sensitive surfaces (such as all high-tech coated surfaces e.g. CRT, LCD and plasma screens) can easily be damaged by a microfiber cloth if it has picked up grit or other abrasive particles during use. One way to minimize the risk of damage to flat surfaces is to use a flat, non-rugged microfiber cloth, as these tend to be less prone to retaining grit.

Rags made of microfiber must only be washed in regular washing detergent, not oily, self-softening, soap-based detergents. Fabric softener must not be used. The oils in the softener and self-softening detergents will clog up the fibers and make them less effective until the oils are washed out.

Insulation

Microfiber materials such as PrimaLoft are used for thermal insulation as a replacement for down feather insulation in sleeping bags and outdoor equipment, due to their better retention of heat when damp or wet.

Basketballs

With microfiber basketballs already popular worldwide and in FIBA, the NBA proposed the use of a microfiber ball for the 2006–07 season. The ball, which is manufactured by Spalding, does not require a "break-in" period of use as leather balls do, and has the ability to absorb water and oils, meaning that sweat from players touching the ball is better absorbed, making the ball less slippery. Over the course of the season, the league received many complaints from players who found that the ball bounced differently from leather balls, and that it left cuts on their hands. On January 1, 2007, the league scrapped the use of all microfiber balls and returned to leather basketballs.

Other

Microfibers used in tablecloths, furniture, and car interiors are designed to repel wetting and consequently are difficult to stain. Microfiber tablecloths will bead liquors until they are removed and are sometimes advertised showing red wine on a white tablecloth that wipes clean with a paper towel. This and the ability to mimic suede economically are common selling points for microfiber upholstery fabrics (e.g., for couches).

Microfibers are used in towels especially those to be used at swimming pools as even a small towel dries the body quickly. They dry quickly and are less prone than cotton towels to become stale if not dried immediately. Microfiber towels need to be soaked in water and pressed before use, as they would otherwise repel water as microfiber tablecloths do.

Microfiber is also used for other applications such as making menstrual pads, cloth diaper inserts, body scrubbers, face mitts, whiteboard cleaners, and various goods that need to absorb water and/or attract small particles.

Environmental and Safety Issues

Microfiber textiles tend to be flammable if manufactured from hydrocarbons (polyester) or carbohydrates (cellulose) and emit toxic gases when burning, more so if aromatic (PET, PS, ABS) or treated with halogenated flame retarders and aromatic dyes. Their polyester and nylon stock are made from petrochemicals, which are not a renewable resource and are not biodegradable. However, if made out of polypropylene, they are recyclable (Prolen).

For most cleaning applications they are designed for repeated use rather than being discarded after use. (An exception is the precise cleaning of optical components where a wet cloth is drawn once across the object and must not be used again as the debris collected and now embedded in the cloth may scratch the optical surface.) In many household cleaning applications (washing floors, furniture, etc.) microfiber cleaning fabrics can be used without detergents or cleaning solutions which would otherwise be needed.

There are environmental concerns about this product entering the oceanic food chain similar to other microplastics. However, no pesticides are used for producing synthetic fibres (in comparison to cotton). If these products are made of polypropylene yarn, the yarn is dope-dyed; i.e. no water is used for dyeing (as with cotton, where thousands of litres of water become contaminated).

Although cotton is more coll crisper than microfiber using as bed linen, microfiber is better for environment due to last longer than cotton, when washing need less water than cotton, dries 3x faster than cotton and microfiber absorb dirt including allergens more effectively than cotton.

References

- Flugel, John Carl (1976) [1930], The Psychology of Clothes, International Psycho-analytical Library, No.18, New York: AMS Press. First published by Hogarth Press, London, ISBN 0-404-14721-6 Alternative ISBN 978-0-404-14721-1

- Bellis, Mary (February 1, 2016). "The History of Clothing – How Did Specific Items of Clothing Develop?". The About Group. Retrieved August 12, 2016.

- "Wilbert L. "Bill" Gore". Plastics Academy Hall of Fame. Archived from the original on April 2, 2015. Retrieved October 22, 2013.

- Hostetter, Kristin. "Is Gore-Tex the Best?". Backpacker. Archived from the original on February 2, 2014. Retrieved January 27, 2014.

- "GORE completes elimination of PFOA from raw material of its functional fabrics" (Press release). W. L. Gore & Associates. January 10, 2014. Retrieved October 7, 2014.

- Richards, Davi (2006-06-02). "Coir is sustainable alternative to peat moss in the garden". Garden Hints. Corvallis, OR: Oregon State University Extension Service. Retrieved 2013-03-06.

- Holt, Stephen (2006-04-12). "Synthetic Clothes Off Limits to Marines Outside Bases in Iraq". US Department of Defense. Retrieved 2012-09-17.

- Ahsan Abdullah (2008-08-22). "AgBioForum 13(3): An Analysis of Bt Cotton Cultivation in Punjab, Pakistan Using the Agriculture Decision Support System (ADSS)". Agbioforum.org. Retrieved 2012-09-17.

- "Accumulation of Microplastic on Shorelines Wolrdwide: Sources and Sinks - Environmental Science & Technology (ACS Publications)". Pubs.acs.org. Retrieved 2012-09-17.

- Dane County Department of Land and Water Resources (2007). Dane County Erosion Control and Stormwater Management Manual (PDF) (Report). Madison, WI. Retrieved 2010-02-09.

- UC Davis Health System: Newroom — UC Davis Pioneers Use Of Microfiber Mops In Hospitals. Ucdmc.ucdavis.edu. Retrieved on 2010-12-01.

Practices of Textile Engineering

This chapter details processes of lace making and embroidery. These require special processes and techniques for manufacture. The chapter also talks about different types of lace like chemical lace, bobbin lace and needle lace. The section on embroidery focuses on machine embroidery.

Lace

Lace is a delicate fabric made of yarn or thread in an open weblike pattern, made by machine or by hand.

Valuable old lace, cut and framed for sale in Bruges, Belgium

Originally linen, silk, gold, or silver threads were used. Now lace is often made with cotton thread, although linen and silk threads are still available. Manufactured lace may be made of synthetic fiber. A few modern artists make lace with a fine copper or silver wire instead of thread.

Etymology

The word lace is from Middle English, from Old French *las*, noose, string, from Vulgar Latin **laceum*, from Latin *laqueus*, noose; probably akin to *lacere*, to entice or ensnare.

Types

There are many types of lace, classified by how they are made. These include:

- Needle lace, such as Venetian Gros Point, is made using a needle and thread. This is the most flexible of the lace-making arts. While some types can be made more quickly than the finest of bobbin laces, others are very time-consuming. Some purists regard needle lace as the height of lace-making. The finest antique needle laces were made from a very fine thread that is not manufactured today.

- Cutwork, or whitework, is lace constructed by removing threads from a woven background, and the remaining threads wrapped or filled with embroidery.

- Bobbin lace, as the name suggests, is made with bobbins and a pillow. The bobbins, turned from wood, bone, or plastic, hold threads which are woven together and held in place with pins stuck in the pattern on the pillow. The pillow contains straw, preferably oat straw or other materials such as sawdust, insulation styrofoam, or ethafoam. Also known as Bone-lace. Chantilly lace is a type of bobbin lace.

- Tape lace makes the tape in the lace as it is worked, or uses a machine- or hand-made textile strip formed into a design, then joined and embellished with needle or bobbin lace.

- Knotted lace includes macramé and tatting. Tatted lace is made with a shuttle or a tatting needle.

- Crocheted lace includes Irish crochet, pineapple crochet, and filet crochet.

- Knitted lace includes Shetland lace, such as the "wedding ring shawl", a lace shawl so fine that it can be pulled through a wedding ring.

- Machine-made lace is any style of lace created or replicated using mechanical means.

- Chemical lace: the stitching area is stitched with embroidery threads that form a continuous motif. Afterwards, the stitching areas are removed and only the embroidery remains. The stitching ground is made of a water-soluble or non-heat-resistant material.

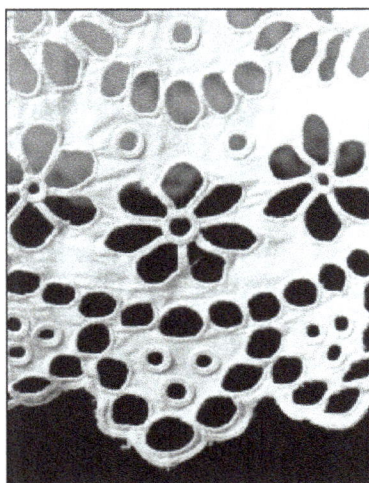

Broderie anglaise, a type of cutwork

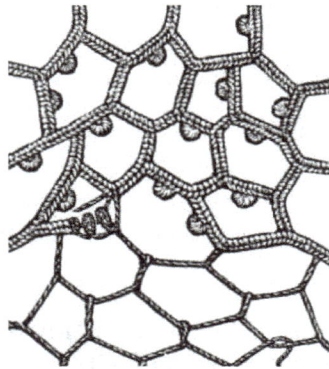

Needle lace, showing button hole stitch

History

The origin of lace is disputed by historians. An Italian claim is a will of 1493 by the Milanese Sforza family. A Flemish claim is lace on the alb of a worshiping priest in a painting about 1485 by Hans Memling. But since lace evolved from other techniques, it is impossible to say that it originated in any one place.

The late 16th century marked the rapid development of lace, both needle lace and bobbin lace became dominant in both fashion as well as home décor. For enhancing the beauty of collars and cuffs, needle lace was embroidered with loops and picots.

Lace was used by clergy of the early Catholic Church as part of vestments in religious ceremonies but did not come into widespread use until the 16th century in the northwestern part of the European continent. The popularity of lace increased rapidly and the cottage industry of lace making spread throughout Europe. In North America in the 19th century, missionaries spread the knowledge of lace making to the Native American tribes. St. John Francis Regis helped many country girls stay away from the cities by establishing them in the lace making and embroidery trade, which is why he became the Patron Saint of lace making.

The English diarist Samuel Pepys often wrote about the lace used for his, his wife's, and his acquaintances' clothing, and on May 10, 1669, noted that he intended to remove the gold lace from the sleeves of his coat "as it is fit [he] should", possibly in order to avoid charges of ostentatious living.

For the industrial revolution, Lace machine.

To date inspiring journals, guilds and foundations show that old techniques with a new twist can challenge young people to create works that can definitely classify as art.

Patrons and Lace Makers

Historical

- Giovanna Dandolo 1457–1462

- Barbara Uthmann 1514–1575

- Morosina Morosini 1545–1614

- Federico de Vinciolo sixteenth-century

Contemporary

- Rosa Elena Egipciaco

Needle Lace

Needle lace (also known as needlelace or needle-made lace or point lace) is a type of lace created using a needle and thread to stitch up hundreds of small stitches to form the lace itself.

Needle lace borders from the Erzgebirge mountains of Germany in 1884, displayed in the Victoria and Albert Museum.

Needle lace, detail

In its purest form the only equipment and materials used are a needle, thread and scissors. This form of lace making originated in Armenia where there is evidence of an Armenian needle lace making tradition dating back to the pre-Christian era. Turkish needle lace is also very popular around the world. This form however arose separately from what is usually termed needle lace and is generally referred to as knotted lace. Such lace is very durable and will not unravel if one or more loops are broken.

Beginning in the 17th century in Italy, a variety of styles developed where the work is started by securing heavier guiding threads onto a stiff background (such as thick paper) with stitches that can later be removed. The work is then built up using a variety of stitches - the most basic being a variety of buttonhole or blanket stitch. When the entire area is covered with the stitching, the stay-stitches are released and the lace comes away from the paper.

Needle lace is also used to create the fillings or insertions in cutwork.

Bobbin Lace

Bobbin lace is a lace textile made by braiding and twisting lengths of thread, which are wound on bobbins to manage them. As the work progresses, the weaving is held in place with pins set in a lace pillow, the placement of the pins usually determined by a pattern or pricking pinned on the pillow.

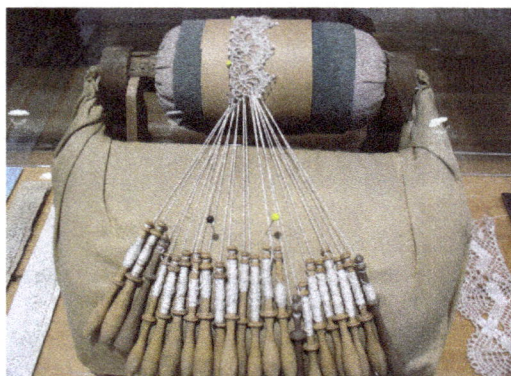

Bobbin lace in progress at the Musée des Ursulines de Québec

Bobbin lace is also known as pillow lace, because it was worked on a pillow, and bone lace, because early bobbins were made of bone or ivory.

Bobbin lace is one of the two major categories of handmade laces, the other being needlelace, derived from earlier cutwork and reticella.

Origin

Early bobbin lace in gold and silver thread, c. 1570.

A will of 1493 by the Milanese Sforza family mentions lace created with twelve bobbins.

Bobbin lace evolved from passementerie or braid-making in 16th-century Italy. Genoa was famous for its braids, hence it is not surprising to find bobbin lace developed in the city. It traveled along with the Spanish troops through Europe. Coarse *passements* of gold and silver-wrapped threads or colored silks gradually became finer, and later bleached linen yarn was used to make both braids and edgings.

The making of bobbin lace was easier to learn than the elaborate cutwork of the 16th century, and the tools and materials for making linen bobbin lace were inexpensive. There was a ready market for bobbin lace of all qualities, and women throughout Europe soon took up the craft which earned a better income than spinning, sewing, weaving or other home-based textile arts. Bobbin lace-making was established in charity schools, almshouses, and convents.

In the 17th century, the textile centers of Flanders and Normandy eclipsed Italy as the premiere sources for fine bobbin lace, but until the coming of mechanization hand-lacemaking continued to be practiced throughout Europe, suffering only in those periods of simplicity when lace itself fell out of fashion.

Structure

Bobbin lace may be made with coarse or fine threads. Traditionally it was made with linen, silk, wool, or, later, cotton threads, or with precious metals. Today it is made with a variety of natural and synthetic fibers and with wire and other filaments.

Elements of bobbin lace may include toile or *toilé* (clothwork), *réseau* (the net-like ground of continuous lace), fillings of part laces, tapes, gimp, picots, tallies, ribs and rolls. Not all styles of bobbin lace include all these elements.

The close up of the back shows the fillings are sewn onto the ribs and tied off	A single plait can choose a clever path to construct a filling with sewings but without tying off	Raised work, a rib on top of the left section, a roll on top of the right section

Traditional Types

Many styles of lace were made in the heyday of lacemaking (approximately the 16th–18th centuries) before machine-made lace became available.

- Classification of traditional styles by technique
  - Continuous bobbin lace also known as: straight lace or fil continu.
    - Mesh grounded lace has motives connected with ground

- too many types to repeat here
- Guipure lace has motifs connected with plaits
- Bedfordshire lace (Beds) – this has flowing lines and picots (to foil the lace machines)
- Cluny lace – has radiating long, thin leaves, called wheatears
- Maltese lace – often has the 8 pointed Maltese cross as part of the pattern
- Yak lace – made of wool
- Genoese lace – usually a geometric design
  - Part lace
- Honiton lace – very fine English lace with many flowers
- Rosaline Perlée – a mixed lace, but mainly bobbin lace
- Bruges lace – assembled from leaves scrolls and open flowers
- Brussels lace – Point d'Angleterre, Point plat appliqué, Point Duchesse
  - Bobbin tape lace sometimes categorized as part lace
- Russian lace
- Idrija lace
- Schneeberg lace – since about 1910
- Milanese lace
- Hinojosa lace
- Peasant lace

Contemporary Laces

Contemporary handmade woollen bobbin lace articles, Wool Expo, Armidale NSW.
Pale green lace is made of 2 ply wool.

The advent of machine-made lace at first pushed lace-makers into more complicated designs beyond the capabilities of early machines, then simpler designs so they could compete on price, and finally pushed them out of business almost entirely.

The resurgence of lace-making is a recent phenomenon and is mostly done as a hobby. Lacemaking groups still meet in regions as varied as Devonshire, England and Orange County, California. In the European towns where lace was once a major industry, especially in Belgium, England, Spain (Camarinas), northern and centre Portugal and France, lacemakers still demonstrate the craft and sell their wares, though their customer base has shifted from the wealthy nobility to the curious tourist.

Still new types of lace are being developed such as the 3D Rosalibre and a colored version of Milanese lace by borrowing rolls from Duchesse lace to store various shades and colors. Other artists are giving grounds a major role by distorting and varying stitches, pin distances and thread sizes or colours. The variations are explored by experimentation and mathematics and algorithms. The lace maintaining its shape without stiffening is no longer a requirement. Inspiring journals, guilds and foundations show that old techniques with a new twist can challenge young people to create works that can definitely classify as art. A Dutch design graduate in 2006 discovered bobbin lace was a technique to make a fancy fence. The first fences became museum pieces. The fences are now produced in Bangalore by concrete rebar plaiters.

Tools

The major tools to make bobbin lace are a pillow, bobbins, pins and prickings. The part laces also require a crochet hook, very fine types of lace require very fine hooks. There are different types of pillows and bobbins linked to areas, eras and type of lace.

prickings for various types of lace and a very fine hook

Types of Bobbins

Spangled bobbins

Fig. 18.   Fig. 19.

winding schemes with a single hitch

Hooded bobbins

Types of Pillow

The pillows must be firm, or otherwise the pins will wobble. The pillows were traditionally stuffed with straw, but nowadays polystyrene (styrofoam) is generally used.

An early type of pillow can be seen in The Lace-Maker by Caspar Netscher. The pillow has a wooden frame, and is slightly sloping. The lace-maker rests it on her lap.

The bolster or cylindrical pillow was much cheaper to make as it is just a fabric bag stuffed with straw. It was used in Bedfordshire lace. It needs a stand as it does not have a flat bottom. Usually the bolster had the pattern pinned round the cylinder, so by turning the pillow, the lace could be as long as was needed. However, Maltese lacemakers used the pillow the other way. They had a long thin pillow, which they rested against something. Then they worked the lace down the length of the pillow.

This problem (of the lace needing to be longer than the pillow) is solved in a different way by the roller pillow, which has a small roller, for working the lace, set into a larger area, where the bobbins are laid. This means that the pattern can be pinned round the roller, but the pillow has a flat bottom.

The cheapest modern pillow is domed and made of polystyrene (styrofoam). It is often called a cookie pillow, because of its shape. Another modern pillow is a block pillow, with a frame which holds covered polystyrene blocks. The blocks can be moved around as the lace progresses, to keep the lace being worked on at the centre of the pillow.

by Vasily Tropinin

Modern domed pillow or "cookie pillow"

Block pillow

Chemical Lace

Chemical lace

Chemical Lace (sometimes referred to as Schiffli Lace) is a form of machine-made lace. This method of lace-making is done by embroidering a pattern on a sacrificial fabric that has been chemically treated so as to disintegrate after the pattern has been created. Schiffli machines came into use in the late 19th century. Before that embroidery machines called Swiss Handmachine were used to make chemical lace as well as embroideries.

This embroidery is nowadays typically done on a multi-head or multi-needle Schiffli machine or loom that has a very large, continuous and overlapping embroidery field. The lace pattern is designed such that the embroidery thread creates an interlocking series of threads that will, in essence, become a "stand alone" piece of lace.

After the embroidery is completed the embroidered fabric is immersed in a solution that will not harm the embroidery thread but completely dissolves the sacrificial fabric leaving just the lace.

Utilizing these large machines and this technique a single piece of lace could be, using today's state-of-the-art machines, over 60" wide by 15yards long. In practice, this system is used to produce many smaller items with one setup.

The original composition of the disintegrating "bath" was not very friendly to the environment and has all but ceased to exist in developed countries. However, the practice is still being used to create laces in third world countries. Since the original development of chemical lace, other methods have been developed beyond the chemical washing method described above. This includes the use of base fabrics that are water soluble or that disintegrate under heat. These methods are generally too expensive or impractical for large-scale production. These are typically used by smaller embroidery facilities specializing in targeted markets, home-based businesses, or hobbyists.

Chemical lace can be distinguished from needle lace by a slight fuzziness in the threads.

Embroidery

Embroidery is the handicraft of decorating fabric or other materials with needle and thread or yarn. Embroidery may also incorporate other materials such as metal strips, pearls, beads, quills, and sequins. Today, embroidery is most often seen on caps, hats, coats, blankets, dress shirts, denim, stockings, and golf shirts. Embroidery is available with a wide variety of thread or yarn color.

Exquisite gold embroidery on the *gognots* (apron) of a 19th-century Armenian bridal dress from Akhaltsikhe.

The basic techniques or stitches on surviving examples of the earliest embroidery—chain stitch, buttonhole or blanket stitch, running stitch, satin stitch, cross stitch—remain the fundamental techniques of hand embroidery today.

History

Traditional embroidery in chain stitch on a Kazakh rug, contemporary.

Caucasus embroidery

Origins

The process used to tailor, patch, mend and reinforce cloth fostered the development of sewing techniques, and the decorative possibilities of sewing led to the art of embroidery. Indeed, the remarkable stability of basic embroidery stitches has been noted:

It is a striking fact that in the development of embroidery ... there are no changes of materials or techniques which can be felt or interpreted as advances from a primitive to a later, more refined stage. On the other hand, we often find in early works a technical accomplishment and high standard of craftsmanship rarely attained in later times.

The art of embroidery has been found worldwide and several early examples have been found. Works in China have been dated to the Warring States period (5th–3rd century BC). In a garment from Migration period Sweden, roughly 300–700 AD, the edges of bands of trimming are reinforced with running stitch, back stitch, stem stitch, tailor's buttonhole stitch, and whipstitching, but it is uncertain whether this work simply reinforced the seams or should be interpreted as decorative embroidery.

Applications and Techniques

Depending on time, location and materials available, embroidery could be the domain of a few experts or a widespread, popular technique. This flexibility led to a variety of works, from the royal to the mundane.

Elaborately embroidered clothing, religious objects, and household items often were seen as a mark of wealth and status, as in the case of Opus Anglicanum, a technique used by professional workshops and guilds in medieval England. In 18th century England and its colonies, samplers employing fine silks were produced by the daughters of wealthy families. Embroidery was a skill marking a girl's path into womanhood as well as conveying rank and social standing.

Conversely, embroidery is also a folk art, using materials that were accessible to non-professionals. Examples include Hardanger from Norway, Merezhka from Ukraine, Mountmellick embroidery from Ireland, Nakshi kantha from Bangladesh and West Bengal, and Brazilian embroidery. Many techniques had a practical use such as Sashiko from Japan, which was used as a way to reinforce clothing.

The Islamic World

Morocco fez horse cover metal silver thread 18th - 19th

In the 16th century, in the reign of the Mughal Emperor Akbar, his chronicler Abu al-Fazl ibn Mubarak wrote in the famous Ain-i-Akbari: "His majesty (Akbar) pays much attention to various stuffs; hence Irani, Ottoman, and Mongolian articles of wear are in much abundance especially textiles embroidered in the patterns of *Nakshi*, *Saadi*, *Chikhan*, *Ari*, *Zardozi*, *Wastli*, *Gota* and *Kohra*. The imperial workshops in the towns of Lahore, Agra, Fatehpur and Ahmedabad turn out many masterpieces of workmanship in fabrics, and the figures and patterns, knots and variety of fashions which now prevail astonish even the most experienced travelers. Taste for fine material has since become general, and the drapery of embroidered fabrics used at feasts surpasses every description."

Automation

The development of machine embroidery and its mass production came about in stages in the Industrial Revolution. The earliest machine embroidery used a combination of machine looms and teams of women embroidering the textiles by hand. This was done in France by the mid-1800s. The manufacture of machine-made embroideries in St. Gallen in eastern Switzerland flourished in the latter half of the 19th century.

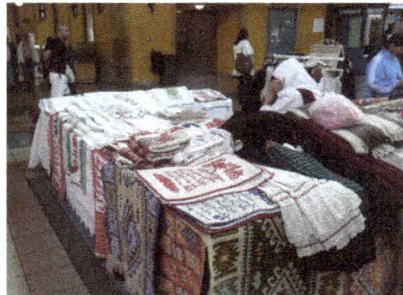
Hand-made embroidery – Székely Land, 2014

Classification

Japanese free embroidery in silk and metal threads, contemporary.

Embroidered Easter eggs. Works by Inna Forostyuk, the folk master from the Luhansk region (Ukraine)

Embroidery can be classified according to whether the design is stitched *on top of* or *through* the foundation fabric, and by the relationship of stitch placement to the fabric.

In free embroidery, designs are applied without regard to the weave of the underlying fabric. Examples include crewel and traditional Chinese and Japanese embroidery.

Cross-stitch counted-thread embroidery. Tea-cloth, Hungary, mid-20th century

Counted-thread embroidery patterns are created by making stitches over a predetermined number of threads in the foundation fabric. Counted-thread embroidery is more easily worked on an even-weave foundation fabric such as embroidery canvas, aida cloth, or specially woven cotton and linen fabrics although non-evenweave linen is used as well. Examples include needlepoint and some forms of blackwork embroidery.

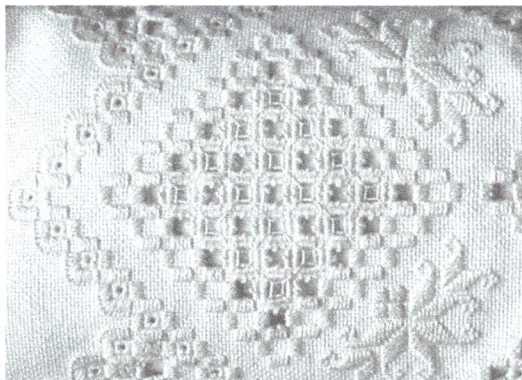

Hardanger, a whitework technique. Contemporary.

In canvas work threads are stitched through a fabric mesh to create a dense pattern that completely covers the foundation fabric. Traditional canvas work such as bargello is a counted-thread technique. Since the 19th century, printed and hand painted canvases, on which the printed or painted image serves as a guide to the placement of the various thread or yarn colors, have eliminated the need for counting threads. These are particularly suited to pictorial rather than geometric designs such as those deriving from the Berlin wool work craze of the early 19th century.

In drawn thread work and cutwork, the foundation fabric is deformed or cut away to create holes that are then embellished with embroidery, often with thread in the same color as the foundation fabric. These techniques are the forerunners of needlelace. When created with white thread on white linen or cotton, this work is collectively referred to as whitework.

Materials

Phulkari from the Punjab region of India. Phulkari embroidery, popular since at least the 15th century, is traditionally done on hand-spun cotton cloth with simple darning stitches using silk floss.

Laid threads, a surface technique in wool on linen. The Bayeux Tapestry, 11th century.

The fabrics and yarns used in traditional embroidery vary from place to place. Wool, linen, and silk have been in use for thousands of years for both fabric and yarn. Today, embroidery thread is manufactured in cotton, rayon, and novelty yarns as well as in traditional wool, linen, and silk. Ribbon embroidery uses narrow ribbon in silk or silk/organza blend ribbon, most commonly to create floral motifs.

Surface embroidery techniques such as chain stitch and couching or laid-work are the most economical of expensive yarns; couching is generally used for goldwork. Canvas work techniques, in which large amounts of yarn are buried on the back of the work, use more materials but provide a sturdier and more substantial finished textile.

In both canvas work and surface embroidery an embroidery hoop or frame can be used to stretch the material and ensure even stitching tension that prevents pattern distortion. Modern canvas work tends to follow symmetrical counted stitching patterns with designs emerging from the repetition of one or just a few similar stitches in a variety of hues. In contrast, many forms of surface embroidery make use of a wide range of stitching patterns in a single piece of work.

Machine

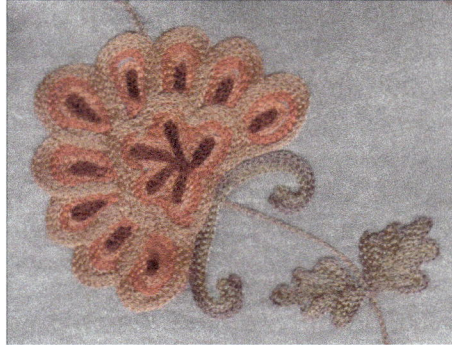

Commercial machine embroidery in chain stitch on a voile curtain, China, early 21st century.

Contemporary embroidery is stitched with a computerized embroidery machine using patterns digitized with embroidery software. In machine embroidery, different types of "fills" add texture and design to the finished work. Machine embroidery is used to add logos and monograms to business shirts or jackets, gifts, and team apparel as well as to decorate household linens, draperies, and decorator fabrics that mimic the elaborate hand embroidery of the past.

There has also been a development in free hand machine embroidery, new machines have been designed that allow for the user to create free-motion embroidery which has its place in textile arts, quilting, dressmaking, home furnishings and more.

Qualifications

City and Guilds qualification in Embroidery allows embroiderers to become recognized for their skill. This qualification also gives them the credibility to teach. For example, the notable textiles artist, Kathleen Laurel Sage- Textiles Artist, began her teaching career by getting the City and Guilds Embroidery 1 and 2 qualifications. She has now gone on to write a book on the subject.

Machine Embroidery

Machine embroidery is an embroidery process whereby a sewing machine or embroidery machine is used to create patterns on textiles. It is used commercially in product branding, corporate advertising, and uniform adornment. Hobbyists also machine embroider for personal sewing and craft projects.

There are multiple types of machine embroidery. These include *free-motion sewing machine embroidery*, this uses a basic zigzag sewing machine. Much commercial embroidery is still done with *link stitch embroidery* the patterns may be manually or automatically controlled. More modern *computerized machine embroidery*, uses an embroidery machine or sewing/embroidery machine that is controlled with a computer that will embroider stored patterns, these may have multiple heads and threads

Free-motion Machine Embroidery

In free-motion machine embroidery, embroidered designs are created by using a basic zigzag sewing machine. As it is used primarily for tailoring, this type of machine lacks the automated features of a specialized machine.

To create free-motion machine embroidery, the embroiderer runs the machine and skillfully moves tightly hooped fabric under the needle to create a design. The operator lowers or covers the "feed dogs" or machine teeth and moves the fabric manually. The operator develops the embroidery manually, using the machine's settings for running stitch and fancier built-in stitches. In this way, the stitches form an image onto a piece of fabric. An embroiderer can produce a filled-in effect by sewing many parallel rows of straight stitching. A machine's zigzag stitch can create thicker lines within a design or be used to create a border. Many quilters and fabric artists use a process called thread drawing (or thread painting) to create embellishments on their projects or to create textile art.

Free-motion machine embroidery can be time-consuming. Since a standard sewing machine has only one needle, the operator must stop and re-thread the machine manually for each subsequent color in a multi-color design. He or she must also manually trim and clean up loose or connecting threads after the design is completed.

As this is a manual process rather than a digital reproduction, any pattern created using free-motion machine embroidery is unique and cannot be exactly reproduced, unlike with computerized embroidery.

With the advent of computerized machine embroidery, the main use of manual machine embroidery is in fiber art and quilting projects. Though some manufacturers still use manual embroidery to embellish garments, many prefer computerized embroidery's ease and reduced costs.

Computerized Machine Embroidery

Most modern embroidery machines are computer controlled and specifically engineered for embroidery. Industrial and commercial embroidery machines and combination sewing-embroidery machines have a hooping or framing system that holds the framed area of fabric taut under the sewing needle and moves it automatically to create a design from a pre-programmed digital embroidery pattern.

Depending on its capabilities, the machine will require varying degrees of user input to read and sew embroidery designs. Sewing-embroidery machines generally have only one needle and require the user to change thread colors during the embroidery process. Multi-needle industrial machines are generally threaded prior to running the design and do not require re-threading. These machines require the user to input the correct color change sequence before beginning to embroider. Some can trim and change colors automatically.

A multi-needle machine may consist of multiple sewing heads, each of which can sew the same design onto a separate garment concurrently. Such a machine might have 20 or more heads, each consisting of 15 or more needles. A head is usually capable of producing many special fabric effects, including satin stitch embroidery, chain stitch embroidery, sequins, appliqué, and cutwork.

History

Before computers were affordable, most embroidery was completed by punching designs on paper tape that then ran through an embroidery machine. One error could ruin an entire design, forcing the creator to start over.

In 1980, Wilcom introduced the first computer graphics embroidery design system to run on a minicomputer. Melco, an international distribution network formed by Randal Melton and Bill Childs, created the first embroidery sample head for use with large Schiffli looms. These looms spanned several feet across and produced lace patches and large embroidery patterns. The sample head allowed embroiderers to avoid manually sewing the design sample and saved production time. Subsequently, it became the first computerized embroidery machine marketed to home sewers.

The economic conditions of the Reagan years, coupled with tax incentives for home businesses, helped propel Melco to the top of the market. At the Show of the Americas in 1980, Melco unveiled the Digitrac, a digitizing system for embroidery machines. The digitized design was composed at six times the size of the embroidered final product. The Digitrac consisted of a small computer, similar in size to a BlackBerry, mounted on an X and Y axis on a large white board. It sold for $30,000. The original single-needle sample head sold for $10,000 and included a 1" paper-tape reader and 2 fonts. The digitizer marked common points in the design to create elaborate fill and satin stitch combinations.

Melco patented the ability to sew circles with a satin stitch, as well as arched lettering generated from a keyboard. An operator digitized the design using similar techniques to punching, transferring the results to a 1" paper tape or later to a floppy disk. This design would then be run on the embroidery machine, which stitched out the pattern. Wilcom enhanced this technology in 1982 with the introduction of the first multi-user system, which allowed more than one person to work on the embroidery process, streamlining production times.

Brother Industries entered the embroidery industry after several computerized embroidery companies contracted it to provide sewing heads. Later, the Japanese company Tajima provided sewing heads that were capable of using multiple threads. Singer failed to remain competitive during this time. Melco was acquired by Saurer in 1989.

The major embroidery machine companies eventually adapted their commercial systems and marketed them to companies such as Janome for home use.

Since the late 1990s, computerized machine embroidery has grown in popularity as costs have fallen for computers, software, and embroidery machines. Many machine manufacturers sell their own lines of embroidery patterns. In addition, many individuals and independent companies also sell embroidery designs, and there are free designs available on the internet.

The Computerized Machine Embroidery Process

Machine embroidery in progress.

Machine Embroidery is not as 'push button' as many people tend to believe and can be a very te-dious process depending on numerous variables. The basic steps for creating embroidery with a computerized embroidery machine are as follows:

- purchase or create (can take hours for the smallest/simplest of designs and the software is costly) a digitized embroidery design file that works with the brand of machine

- edit the design and/or combine with other designs with costly software (optional)

- load the final design file into the embroidery machine after making sure it is the right for-mat and it will fit in the hoop you need

- stabilize the fabric and place it in the machine

- start and monitor the embroidery machine, to change thread colors, rethread machine, troubleshoot problems, etc. and have plenty of needles, bobbins, a can of air (or small air compressor), a small brush, and scissors on hand

- Toss out botched project due to wrong stabilizer for item, machine malfunction, wrong thread, badly digitized design, etc.

- Repeat process from the beginning till its right

Design Files

Digitized embroidery design files can be either purchased or created with industry-specific em-broidery digitizing software. Embroidery file formats broadly fall into two categories. The first, source formats, are specific to the software used to create the design. For these formats, the digi-tizer keeps the original file for the purposes of editing. The second, machine formats, are specific to a particular brand of embroidery machine. Here, the files are available for use with particular embroidery machines and are not easily edited or scaled.

Wilcom developed the .EMB file format which is the designer's file format of choice ensuring op-timum stitch quality. This format stores the true object based properties and means that they can easily resize, re-color, adjust for fabric types, etc with a simple click for each project. Once ready the .EMB is exported into any machine format required for that specific stitch-out.

Embroidery machines generally have one or more machine formats specific to their brand. Howev-er, some formats such as Tajima's .dst, Melco's .exp/.cnd and Barudan's .fdr have become so prev-alent that they have effectively become industry standards and are often supported by machines built by rival companies.

Machine formats generally contain primarily stitch data (offsets) and machine functions (trims, jumps, etc.) and are thus not easily scaled or edited without extensive manual work.

Many embroidery designs can be downloaded in popular machine formats from embroidery web sites. However, since not all designs are available for every machine's specific format, some ma-chine embroiderers use conversion programs to convert from one machine's format file to another, with various degrees of reliability.

A person who creates a design is known as an embroidery digitizer or puncher. A digitizer uses software to create an object-based embroidery design, which can be easily reshaped and edited. These files retain important information such as object outlines, thread colors, and original artwork used to punch the designs. When the file is converted to a stitch file, it loses much of this information, rendering editing difficult or impossible.

Software vendors often advertise auto-punching or auto-digitizing capabilities. However, if high quality embroidery is essential, then industry experts highly recommend either purchasing solid designs from reputable digitizers or obtaining training on solid digitization techniques.

Editing Designs

Once a design has been digitized, an embroiderer can use software to edit it or combine it with other designs. Most embroidery programs allow the user to rotate, scale, move, stretch, distort, split, crop, or duplicate the design in an endless pattern. Most software allows the user to add text quickly and easily. Often the colors of the design can be changed, made monochrome, or re-sorted. More sophisticated packages allow the user to edit, add, or remove individual stitches. Some embroidery machines have rudimentary built-in design editing features.

Loading the Design

After editing the final design, the file is loaded into the embroidery machine. Different machines require different formats that are proprietary to that company. Common design file formats for the home and hobby market include .ART, .HUS, .JEF, .PES, .SEW, and .VIP. Embroidery patterns can be transferred to the computerized embroidery machines through cables, CDs, floppy disks, USB interfaces, or special cards that resemble flash or compact cards.

Stabilizing the Fabric

To prevent wrinkles and other problems, the fabric must be stabilized. The method of stabilizing depends on the type of machine, the fabric type, and the design density. For example, knits and large designs typically require firm stabilization. There are many methods for stabilizing fabric, but most often one or more additional pieces of material called stabilizers or interfacing are added beneath or on top of the fabric, or both. Stabilizer types include cut-away, tear-away, solvy water-soluble, heat-n-gone, filmoplast, and open mesh, sometimes in various combinations.

For embroidered wearable items, the fabric is placed in a hoop. This is then attached to the machine . An X and Y drive mechanism moves the hoop under the needle following the design coordinates created when the design was digitized for embroidery.

Learn more on choosing the right stabilizer for your embroidery job here

Embroidering the Design

Finally, the embroidery machine is started and monitored. For commercial machines, this process is more automated than for the home machines. Many designs require more than one color and may involve additional processing for appliqués, foam, or other special effects. Since home machines have only one needle, every color change requires the user to cut the thread and change the color manually. In addition, most designs have one or more jumps that need to be cut. Depending on the quality and size of the design, sewing a design file can require anywhere from a few minutes to over half a day!

Embroidery Machines

Not all machines are solely used for embroidery; some are also used for sewing. Some of the more advanced features becoming available include a large color touchscreen, a USB interface, auto threading, built-in design editing software, embroidery adviser software, and design file storage systems. Commercial embroidery machines can be purchased with a set number of needle colors per head(1, 2, 3, 4, 6, 12, 15,18 or more colors). Industrial embroidery machines are available with 1 to 56 heads.

Commercial and Contract Embroidery Factories

Factories can have a few small machines or many large machines, or any combination of machines. Contract embroidery is done on goods that the customer supplies to the embroidery house and is limited to the trade, "ASI" and marketing firms use these services almost exclusively. A company offering contract embroidery sews designs onto wearable items for brokers, other embroiderers, specialty firms, and screen printers at a wholesale rates. The customer of a contract embroiderer usually supplies the items to the factory and only pays for the embroidery service.

Commercial embroiderers, and some contract embroiderers, offer their services to the public, and can supply the wearable items, and usually have a vast collection of stock designs and text available, Keeping up with current market trends, and offering names and personalization as well as designs for embroidery.

Other Supplies

Almost any type of fabric can be embroidered, given the proper stabilizer. Base materials include paper, fabric, and lightweight balsa wood.

Machine embroidery commonly uses polyester, rayon, or metallic embroidery thread, though other thread types are available. 40 wt thread is the most commonly used embroidery thread weight. Bobbin thread is usually either 60 wt or 90 wt. The quality of thread used can greatly affect the number of thread breaks and other embroidery problems. Polyester thread is generally more color-safe and durable. High quality embroidery thread is produced by Exquisite, Gunold, Madeira, and Robison-Anton.

Other associated costs are thread, stabilizer, purchased designs, needles, bobbins, and other miscellaneous tools and supplies.

Embroidery Glossary

A more thorough list of applicable terminologies is available at http://abcoln.com/glossary.php.

Appliqué

> French term meaning applying, usually by sewing, one piece of fabric to the surface of an-

other. A cut piece of material stitched to another adds dimension and texture and reduces the stitch count.

Backer/Stabilizer

Backing and stabilizer are often used interchangeably to refer to materials, generally non-woven textiles, which are placed inside or under the item to be embroidered. The backing provides support and stability to the garment which will improve the quality of the finished embroidered product. Backings come primarily in two types: cutaway and tear-away. With cutaway, the excess backing is cut with a pair of scissors. With tear-away, the excess is torn away after the item is embroidered. Additional types of stabilizer can be dissolved by water or heat.

Bobbin

A small spool of thread inside the rotary hook housing of a sewing machine. The bobbin thread forms the stitches on the underside of the garment. Bobbin thread holds the top embroidery thread to the garment. The bobbin on an embroidery machine works in the same manner and for the same purpose as on a standard sewing machine.

Digitize

The computerized technique of turning a design image into an embroidery program. Special software is used to create plotting commands for the embroidery machine. The commands are transferred to the machine's logic head by a designated embroidery language.

Fill Stitch

Fill stitches are a series of running stitches sewn closely together to form broad areas of embroidery with varying patterns and stitch directions.

Hoop

A clamping device used to hold the backing and fabric in place in the machine.

Running Stitch

One straight line of stitches, often used for fine details, outlining, and underlay.

Satin Stitch

Also known as zigzag stitch, a satin stitch is a line, border or edge produced by thread being alternately stitched to either side of a baseline. Satin stitches are generally limited to a maximum of 1/2" in stitch length before some alternate technique must be used, such as split stitching or fill stitching.

Underlay

A stabilizing pattern of embroidery which, if used, precedes the main body of satin or fill stitching. It consists of one or a combination of running stitches for centering, edging, paralleling, or zigzagging the design area. A money and time saving technique is to use, instead of a large amount of embroidery thread for underlay, a fancy specialty stitch saver patch material that simulates underlay.

References

- Levey, S. M.; D. King (1993). The Victoria and Albert Museum's Textile Collection Vol. 3: Embroidery in Britain from 1200 to 1750. Victoria and Albert Museum. ISBN 1-85177-126-3.

- Quinault, Marie-Jo (2003). Filet Lace, Introduction to the Linen Stitch—Instruction book to learn How to do Embroidery on a Knotted Net – FILET LACE BY THE SEA. Trafford Publishing. ISBN 1-4120-1549-9.

- Netherton, Robin, and Gale R. Owen-Crocker, editors, (2005). Medieval Clothing and Textiles, Volume 1. Boydell Press. ISBN 1-84383-123-6.

- Irvine, Veronika; Ruskey, Frank (2014). "Developing a Mathematical Model for Bobbin Lace". Journal of Mathematics and the Arts. 8 (3-4): 95–110. arXiv:1406.1532.

- "History of Lace | Lace Trends | Lace Spreads". Decoratingwithlaceoutlet.com. Archived from the original on 8 March 2014. Retrieved 2012-09-11.

- "Lacemaking: Associations and Guilds". Fibre Arts Online Web. Archived from the original on February 3, 2012. Retrieved 8 August 2011.

Permissions

We would like to thank the editorial team for lending their expertise to make the book truly unique. They have played a crucial role in the development of this book. Without their invaluable contributions this book wouldn't have been possible. They have made vital efforts to compile up to date information on the varied aspects of this subject to make this book a valuable addition to the collection of many professionals and students.

This book was conceptualized with the vision of imparting up-to-date and integrated information in this field. To ensure the same, a matchless editorial board was set up. Every individual on the board went through rigorous rounds of assessment to prove their worth. After which they invested a large part of their time researching and compiling the most relevant data for our readers.

The editorial board has been involved in producing this book since its inception. They have spent rigorous hours researching and exploring the diverse topics which have resulted in the successful publishing of this book. They have passed on their knowledge of decades through this book. To expedite this challenging task, the publisher supported the team at every step. A small team of assistant editors was also appointed to further simplify the editing procedure and attain best results for the readers.

Apart from the editorial board, the designing team has also invested a significant amount of their time in understanding the subject and creating the most relevant covers. They scrutinized every image to scout for the most suitable representation of the subject and create an appropriate cover for the book.

The publishing team has been an ardent support to the editorial, designing and production team. Their endless efforts to recruit the best for this project, has resulted in the accomplishment of this book. They are a veteran in the field of academics and their pool of knowledge is as vast as their experience in printing. Their expertise and guidance has proved useful at every step. Their uncompromising quality standards have made this book an exceptional effort. Their encouragement from time to time has been an inspiration for everyone.

The publisher and the editorial board hope that this book will prove to be a valuable piece of knowledge for students, practitioners and scholars across the globe.

Index

www.ingramcontent.com/pod-product-compliance
Lightning Source LLC
Chambersburg PA
CBHW08202119O326
41458CB00010B/3237